ABOUT THE AUTHORS

John Allen is Executive Chairman of Space Biospheres Ventures, an extraordinary private venture project to create the first materially-enclosed, energetically and informationally-open biosphere, "Biosphere II". A renowned pioneer in ecotechnics, total systems entrepeneurial management, metallurgy and the philosophy of systems theory, John graduated from Colorado School of Mines and, as a Baker Scholar, from Harvard Business School. He has worked on regional development projects in Iran and West Africa with David Lillienthal's Development Resources Corporation, headed a special metals team with Allegheny-Ludlum, and was a founder and serves as consultant to the Institute of Ecotechnics. John has traveled extensively exploring the wealth of ecology and culture on the planet. He served as scientific editor of *The Biosphere Catalogue*, and authored *Succeed: A Handbook on Structuring Managerial Thought*.

Mark Nelson is Chairman and C.E.O. of the Institute of Ecotechnics, a London-based ecological development institute, which conceived, and manages the research and development of the Space Biospheres project. The Institute of Ecotechnics has set up demonstration cost-effective projects in some of the most challenging and exotic biomes of the biosphere -- rainforest, high altitude desert, semi-arid Australian Outback savannah, Asian and Western city centers. A summa cum laude graduate in philosophy from Dartmouth College and chief agronomist with the Institute of Ecotechnics, Mark is a director of the Savannah Systems desertification-control tropical plantation in Western Australia, and operates experimental arid land rain-catchment orchards in the Southwest of the United States. Mark authored a presentation of the Institute of Ecotechnics' conceptual model for ecotechnic management -- harmonizing man, culture and technics with the biosphere -- in *Man, Earth and the Challenges*, and was a contributing editor for *The Biosphere Catalogue*.

Space Biospheres

by
John Allen and Mark Nelson

SP
Synergetic Press, Inc.

Published by Synergetic Press, Inc.
Post Office Box 689, Oracle, Arizona 85623
24 Old Gloucester Street, London WC1 3AL

Copyright © 1986 by John Allen and Mark Nelson.

All rights reserved. No part of this publication may be reproduced, stored in a retrieval system, or transmitted, in any form or by any means, electronic, mechanical, photocopying, recording, or otherwise, without the prior permission of the publisher.

Cover and book design by Kathleen Dyhr.
Typesetting by Synergetic Press on Studio Software.

ISBN 0-907791-050

Printed in the United States of America by Arizona Lithographers.

Contents

Introduction	*1*
Chapter 1 - Biosphere I	*3*
Chapter 2 - Biosphere II	*45*
Chapter 3 - On the Classification of Biospheres	*67*
Chapter 4 - Mars Settlement	*73*
Chapter 5 - Challenges and Opportunities	*86*

Introduction

Biospherics, the integrative science of the life sciences as astronautics is the integrative science of the physical sciences, has implications far beyond the profound possibilities opened up for the understanding of Biosphere I (the biosphere which Homo sapiens now inhabit on planet Earth). Namely, biospherics opens up, together with astronautics, the ecotechnical possibilities, even the historic imperative, to expand Earthlife into the solar system and beyond that to the stars and then in time's good opportunity to the galaxies, perhaps in association with biospheres from other origins.

This book, *Space Biospheres,* for the first time structures the grand geobioconceptual work of Vernadsky and his Russian successors, with the atmospheric-microbiological achievements of Lovelock, Margulis, Folsome, and with the Institute of Ecotechnics' series of biomic and biospheric projects and modelling. The vast panorama of the extraordinary cultural agenda now opening to humanity through the synergy of biospherics and astronautics is the exploration we shall undertake.

The book commences with a presentation of the basic integrative model of Biosphere I; goes on to consider the modelling of the Biosphere II Project now being built by Space Biospheres Ventures near Tucson, Arizona; advances to make a classification of all possible biospheres, an agenda of both creation and study; then considers a practical Mars settlement that takes biospherics and the astrogeologist's studies as its basis for design with the aim of permanent habitation of that planet; and concludes with a chapter on the astonishing challenges and opportunity suddenly available to Homo sapiens, the challenge being to create cultures of succeed to replace the cultures of scraping-by, to create the basis for a genuine nomics, the integrative science of behavioral sciences.

Our destiny, as always, waits before us -- the open road. We must open our eyes, concentrate our attention, and taking the first step in the right direction, become masters and enjoyers of the great adventure.

1

Biosphere I

Although the term *biosphere* is increasingly used, it has been inadequately defined by many as "the thin layer of life on the surface of Earth". This definition scarcely hints at the monumental scale and many-leveled complexity of its activities shaping the global environment by producing the vast movements of matter which sustain evolving life on our planet and indeed mold the crust and atmosphere of the planet itself.

The study of Biosphere I, Earth's biosphere, the only presently known biosphere, challenges many assumptions implicit in previous views of the phenomena of life as "fragile" or a helpless passenger on "Spaceship Earth". The newly emerging science of biospherics reveals Biosphere I to be even more than the powerful and ancient planetary-wide system which has survived vast geologic perturbations, catastrophic mass extinctions and maintained a continuous presence on Earth for over 3.5 billion years, with a future perhaps as long as its past. Biosphere I is also the major geologic force shaping the equilibrosphere (gas, solid and liquid) of the planet and supplies most of the free energy that powers our technosphere, offering poets, scientists and ordinary citizen alike that creative inspiration Jack London named "the call of the wild."

A biosphere is a stable, complex, adaptive, evolving life system with the potential of operating in the right conditions as the major geological force transforming a planet's crust and as the source of sufficient free energy to power the start-up of a technosphere.

The Scale of Biosphere I

Scale has been called the most difficult, yet the most important, of all ideas to grasp in order to succeed in understanding any given phenomenon. The next sections from Mass to Energy give an approach to Biosphere I's scale of operations.

Mass. The current biomass of the biosphere approximates 2.5 trillion dry tons, green plants making up some 99 percent of the total. The net annual productivity -- new organic material -- is under ten percent, and is estimated at between 150 and 200 billion tons of dry organic matter per year. This means that the one percent of the planetary biomass which are animals and saprophytes (decomposers of organic material) annually consume approximately ten times their own weight of 20 billion tons.

Although land photosynthesizers outweigh their ocean counterparts at least a hundred-fold, the marine organisms produce some 25-40 percent of the annual total. The land producers are dominated by slow-growing trees, while the ocean phytomass is principally free-floating and rapidly reproducing plankton, a microscopic algae. The microbes are the unsung heroes of the productivity of Biosphere I.

Active Volume of Life. Nearly all of the life in Biosphere I is concentrated within several hundred feet above and below the surface of the Earth. Life ranges from sediment layers below the deepest levels of the ocean, approximately 38,000 feet below sea level, to the 29,000 feet above sea level peaks of the highest mountains, although some viable dormant microorganisms have been found at an altitude of nearly 50 miles. In relation to the planetary radius of around 4,000 miles and atmosphere of 180 miles, the thin layer of the living biosphere is concentrated in less than one percent of the planetary volume.

Geological Volume of Life. But to understand the full import of the ceaseless activity of the biosphere we must include the continual depositing of a portion of this vast production outside the food-chain cycle in the form of buried dead organisms, waste products, and byproducts. This part of the production is not recycled but accumulates in spheres surrounding the biosphere. V.I. Vernadsky, the Russian geologist who laid the basis for the scientific understanding of Earth's biosphere, summarized the impact of the biosphere as "the most powerful geological force, growing with time ... [which] not only creates the whole picture of our natural surroundings but penetrates into the deepest and most grandiose processes in the Earth's crust."[1]

These concentric spheres which the biosphere creates around itself have been variously named. "Metabiosphere" denotes the upper crust of Earth, where the deposits of past biospheres have shaped the formation of sedimentary rocks, mineral deposits, soils etc. It has been estimated that two to eight percent of the biosphere's annual product becomes an "output to geology", deposited in the Earth's crust or added to its atmosphere.

"Parabiosphere" (alongside the biosphere) designates the lower atmosphere where life-forms, the "aeroplankton," can survive but only in a dormant or sporelike state. "Apobiosphere" refers to the upper atmosphere (above 50 miles) where life-forms cannot exist. Recent work by microbiologists have demonstrated that the prokaryotic microbes, the most ancient life of the biosphere, whose cells lack a nucleus, "produce and remove all of the major reactive gases in the Earth's atmosphere: nitrogen, nitrous oxide, oxygen, carbon dioxide, carbon monoxide, several sulfur-containing gases, hydrogen, methane and ammonia, among others".[2] The atmosphere of Biosphere I is life-created and life-maintained. The geological volume of the biosphere occupies over four percent of planet Earth's volume.

Energy. In tropical and temperate regions, solar energy entering the Earth's envelope is estimated to average 700 watts/square meter, of which, after losses to the atmosphere, clouds, etc., some 150 - 250 watts/square meter are available to photosynthesizers who finally utilize from 0.2 to 1.0 percent, 0.5 to 3 watts/square meter. The total energy of biospheric photosynthesis is 40×10^{12} watts, equivalent to 1.2×10^{18} BTU per year, or more than four times the total energy produced by man's technics in 1979, 2.7×10^{17} BTU.

The Evolutionary History of Biosphere I

The evolutionary history of Biosphere I can be understood by studying, after grasping scale, its basic continuity, its growth, its discontinuities in meeting the challenges of catastrophe, its cycles, its diversity, and its force.

Basic Continuity. The oldest rocks geologists have found with fossils resembling microbial life-forms, similar to bacteria, are three and a half billion years old. Life undoubtedly, of course, was present before this time. Studies of the DNA molecule, the basis of genetic coding and inheritance, have revealed that all known Earth life-forms have the same chemical structure, evidence that there has never been a complete break in the continuity of life in our biosphere. An astonishing fact given that conditions on the early Earth were quite different from those at present: a reducing atmosphere probably dominated by methane and carbon dioxide, certainly with little or no free oxygen, solar output at least a third less than at present, and subject to full dosages of deadly ultraviolet radiation since the protective ozone layer did not yet exist.

Growth. Vernadsky identified the photosynthesizers as the key force in biospheric operation for they establish the bond between our planet and the cosmos by utilizing the energy of sunlight for making chemical compounds. "The substance of this boundary region, the biosphere, becomes active under the influence of the stream of energy. It accumulates and distributes the energy received and finally transforms it into free energy in the biosphere." The invention of chlorophyll enabled the early biosphere to use incoming solar energy to utilize hydrogen gas, hydrogen sulfide and later the water molecule by combining them with carbon dioxide from the atmosphere to produce sugars, the simplest of hydrocarbons, some of which can then be built into the proteins indispensable for life.

This basic equation can be expressed:

$$6\,H_2A + 6\,CO_2 + ENERGY = C_6H_{12}O_6 + 6\,A_2$$

where A can be nothing or S, sulfur as in the case of the first photosynthesizers, or O, oxygen when later life developed the capacity to use water as a raw material for the process.

The early biosphere operated without free gaseous oxygen. Recent discoveries by microbiologists have established that single-celled prokaryotes (organisms lacking a nucleus) were the Earth's primordial producers operating through photosynthesis and chemosynthesis. The metabolically eclectic prokaryotes formed the biospheric population for some two billion years, growing in power until they changed the carbon dioxide atmosphere into the more potent oxygen/nitrogen mix and then by controlling this atmospheric composition by complex feedback loops, shaped its own destiny.

The implications of this "longest-running show on the planet" for the biosphere's ability to adapt to and mold its environment successfully is underlined by Clair Folsome, the closed ecosystems pioneer from the University of Hawaii. He points out that microbes enjoy an evolutionary lineage dating back over three billion years (as contrasted with three million for Homo sapiens), and a reproduction time of one to ten hours, compared with a human's 20 years. This means microbes have had some 175 million times as many generations than we to evolve ever finer adaptive mechanisms to become "biochemical and metabolic wizards". Microbes have been found in boiling hot springs, in strong acids, in brackish water ten times more saline than seawater, at soil depths exceeding ten thousand feet, and remained viable after years of freezing. Recently they have been found living in the cooling pipes of nuclear energy plants, and dormant but alive on the Moon in materials left by earlier space missions.

After building a biosphere similar in mass action to the biosphere today, sections of the victorious prokaryotes made a symbiotic coalescence as eucaryotes and with their new internalized power of chloroplasts and mitochondria new econiches were filled, animals, plants, and fungi evolved, until nearly every piece of the surface of planet Earth was colonized by someone.

Large-Scale Discontinuities in Catastrophes. Biologists have found records of at least five mass extinctions of life in the fossils of the Phanerozoic Eon, the most well-known and recent one including the disappearance of the dinosaurs 66 million years ago. The biosphere may be immortal within certain unknown limits upon planet Earth, but its families, genera, and species are not. Estimates of current species, including a majority as yet undiscovered, range from ten to perhaps thirty million as compared with a total estimated five hundred million to one billion species which have lived on the planet. The mean lifetime for a species is about 15 million years.

At times of mass extinctions up to 96 per cent of the

biosphere's marine animal species have been eliminated and half of all taxonomic families. While individual species losses are probably a function of localized factors such as alteration of or increased competition for econiche, mass extinction events relate to global crises, triggered by geological forces (vulcanism, sea level changes, continental drift) or cosmic ones such as climatic perturbations causing glaciation or the impact of comets or asteroids. A recent NASA study noted "the main effect of mass extinctions is to reset evolutionary systems ... which [itself] may be a vital evolutionary process necessary for complex life as we know it ... [preventing] a steady state of evolution because evolutionary change within [species] lineages are known to be slow and relatively inconsequential".[3]

One of the major biospheric catastrophes before the Phanerozoic Eon was itself life-precipitated: the creation of an oxidizing atmosphere, some 1.5 billion years ago. This highly active atmosphere triggered a crisis which undoubtedly led to the extinction of large numbers of anaerobic life-forms which had previously composed the biosphere, and which forced the remainder to their present remnant habitats in the soil and deeper underground, in the off-shore and ocean bottoms, and in the digestive "bottoms" of terrestrial animals. This "environmental crisis" also led to the previously mentioned evolution of the eucaryotes, nucleated-cells, probably through the union of symbiotic prokaryotic cells. Some of the eucaryotes, at first single-celled like the vast majority of the prokaryotes, would later, thanks to their more complex structures, be able to evolve into the complex, multi-cellular fungi, plants and animals.

Under the near breaking-point stimulus of necessity, eucaryotic life-forms evolved the capacity to use the free gaseous oxygen, to emerge from the waters of the planet to occupy its land surface, for through the operation of physical processes once there was but one percent of the current oxygen levels in the atmosphere, an ozone layer formed protecting surface life from deadly short-wave ultraviolet radiation.

This pattern recurs in Earth's evolutionary history, catastrophic change of boundary conditions producing first a drastic toll in existing lines to be followed by transformation and a dazzling burst of speciation.

Gas Cycles. Though the fossil and chemical evidence records Biosphere I's success in overcoming cosmic, geologic and life-induced catastrophes, a look at our neighboring planets, Mars and Venus, shows us two failure modes, that of freezing and that of boiling.

Mars appears to have had a dense atmosphere at an earlier era, with all the signs of liquid flowing water, but with only 38 percent of Earth's gravity, could not hold its volcanic outgassings at a level higher than one percent of Earth's. This fossil atmosphere does not retain sufficient pressure and heat for water to exist in the liquid phase.

Venus, on the other hand, demonstrates positive feedback, or what can result from a runaway "greenhouse effect". Endowed with a thick atmosphere of carbon dioxide which produces surface pressures 100 times that of ours, its surface temperatures average 900 degrees F (480 degrees C).

Biosphere I has avoided both the Martian and Venusian extremes by developing an elaborate series of homeostatic cycles, over varying time-intervals, to moderate fluctuations and ensure vital life elements remain available. In addition, its vast mass, distributed patchily, forms great buffers or refugia from which it can make comebacks. Besides these virtues, the biosphere developed on, as some scientists are fond of pointing out, a planet farther from the Sun than Venus and closer to the Sun than Mars. Nonetheless, the properties existing on the planet Earth are quite different from those predicted by physical equations alone, and the maintaining of these differences must be directly attributed to the recycling powers of the biosphere.

Water in many ways is the keystone of the biosphere, (Vernadsky liked the definition of life as "animated water") constituting some 70 percent of both organic tissue and planetary surface, and being a medium of transport for many of the life elements. Some 97 percent of the Earth's water is saltwater, contained in its seas and oceans. Of the remaining three percent which is fresh water, three-quarters are tied up in ice-caps and glaciers. An extremely small quantity is present in the atmosphere as water vapor, yet as Leonardo da Vinci intuited "water drives nature", being crucial in retaining the heat produced by incident solar energy and energizing the planetary weather patterns.

Solar energy drives the driver, directly transporting water evaporation from land and ocean, and indirectly moving it through the transpiration and photosynthesis of microbes and green plants. Water contained in living biomass totals five times that contained in all the rivers of planet Earth, and the entire water of the planet passes through the life in the biosphere every two million years. To produce 20 fresh-weight tons of crop, or five tons of dry weight, some 2,000 tons of water will have been drawn up by the plants' roots and transpired.

While the water cycle uses the ocean as its principal reservoir, the carbon cycle utilizes the storage of huge inventories deposited in sedimentary rock formations in the Earth. Less than one per cent of the total carbon stock is in rapid biospheric circulation between atmosphere, soil, ocean and living organisms, mainly in the form of carbon dioxide and in the tissues of organisms. The bulk is locked away in the form of carbonate rocks, limestone and fossil fuels (coal, oil and natural gas) which required millions of years to accumulate.

The interplay of the water, carbon and oxygen cycles is rather an intricate balance -- for it now appears that the prime source of new oxygen to replenish that lost in the oxidation of inorganic rocks is the burying of a certain percentage of reduced organic material annually. Each atom of reduced carbon added to the Earth's metabiosphere, under the continents or deposited on the ocean bottom, releases two atoms of oxygen. The quantities of oxygen involved in photosynthetic and respiratory functions (in the form of carbon dioxide) are now thought to be roughly equal. Thus if the biosphere failed to continue removing a small percentage (0.1 percent) of new organic matter annually, the oxygen in the atmosphere would disappear over several thousand years; while if too much were removed, the atmosphere would quickly rise from its present value of 21 percent oxygen to levels above 25 percent where fires would be impossible to control.

James Lovelock, the atmospheric scientist, and the microbiologist Lynn Margulis, who have led the recent rethinking of the importance of the autonomy of biospheric processes (what they call "Gaia") to the operation of our planetary environment, have suggested that the key regulating biospheric agents are the anaerobic methane-producing bacteria whose output of around a billion tons of methane annually to the atmosphere serves as a check preventing increased levels of free atmospheric oxygen.

The cycling of nitrogen also involves all regions of the megabiosphere, but unlike water which uses the ocean as its great reservoir, carbon which uses sedimentary deposits, and oxygen which rebuilds itself from life, nitrogen's main reservoir is in the atmosphere. Nitrogen can be converted to available chemical form through inorganic processes like lightning but mainly as a result of the action of special types of soil and marine microorganisms which "fix" some 90 percent of the total. In many cases (the Leguminosae especially) these microbes are symbiotically connected with plant roots. This first step of making nitrogen available to life (nitrogen fixation) requires an expenditure of energy, whereas all the other cycling steps where nitrogen is built into organic compounds and finally oxidized (de-nitrified) release energy.

The microbial by-products of nitrogen oxidation are all gases -- N_2 (which perhaps plays the role of pressure maintainer and oxidation [fire] dampener in the atmosphere); NO_2 (thought to check excessive ozone formation, and an oxygen contributor balancing methane's reducing role); and ammonia, NH_3 (which controls the pH of rainfall, which would otherwise become dangerously acidic). Annual atmospheric addition of these are on the order of 300, 100, and 300 million tons, respectively. Human manufacture of nitrogenous fertilizer was 45 billion tons in 1978. The disruption of lake and stream ecosystems from nitrogen deposit (eutrophication) may be one "reason" why the biosphere acts to keep its reservoir in the atmosphere rather than the ocean, where its presence would also pose salinity problems for marine organisms.[4]

Mineral Cycles. The cycling of virtually every naturally-occurring element of the periodic table is in some way controlled by biospheric processes. Recent discoveries have linked to life processes even such rare metals as vanadium -- primitive marine chordates like ascidians contain up to 15 per cent of it in their bloodstream, and are even being cultivated in Japanese coastal waters as "living vanadium mines" - and nickel, which constitutes one percent of the leaves of a New Zealand shrub.[5]

Some geologists now recognize that all sedimentary rock deposition is to some extent biologically mediated, and the biospheric functions in the transport and concentration of mineral elements are increasingly recognized as a major force in crustal geology. Andrey Lapo, the Russian geologist, citing recent work in the field, pointed out that "the vegetational cover of our planet annually concentrates mineral matter in amounts comparable to ... their reserves in the lithosphere."[6]

Some figures for annual concentration in photosynthesis (in tons):

phosphorus	1 billion
iron	100 million
manganese, copper, zinc	10 million each
nickel	1 million
cobalt and chromium	100,000 each

Elephant grass in African savannahs extracts from each hectare (2.5 acres) 250 kilograms of silica, 80 kilograms of alkalis; and rainforest vegetation can extract up to nine tons of silica annually per hectare. Microbes and fungi in the top six inches of soil weigh some two tons per acre. As microbes have evolved specialist types which can metabolize virtually any compound, man puts them to direct mining use in the extraction of copper, uranium, zinc and even arsenic; and is developing microbial processes for leaching lead, nickel, cobalt, molybdenum, cadmium and titanium.

As a final illustration we can consider the soil and mud-eaters of land and sea. Darwin, in his classic study of the earthworm, calculated that on fertile English soils, with two million worms per acre, the top meter of soil passes through their digestive systems every two hundred years, and their excrement produces an inch of new topsoil every 50 years. In the nutrient-rich coastal waters off the California coast, marine polychaetes annually ingest 1.5 meters of bottom sediment.

This biospheric function is not limited to autotrophic organisms, though bacteria have been called "not the autocratic creators of mineral deposits, but their natural concentration specialists", their concentrations of various elements reaching levels one million times above the surrounding environment. The heterotrophs continue the process, though we have learned this with dismay in the case of pesticides or radioactive isotopes such as strontium-90, iodine-131. Initially low concentrations build up trophic level by level until at the head of the food chain they become quite poisonous. The concentration factor (above the environment) of strontium-90 in a Canadian lake went from one in the water, to 200 in bottom sediments, 300 in aquatic plants, 1,000 in minnows, 3,000 in perch bone, 3,900 in muskrat bone.[7]

The ceaseless metabolism of living organisms and the movements of animals, including migration, serve the cyclic function of transporting materials against an energy-gradient. The biosphere is a vast negentropic, anti-gravitational apparatus.

Daily and Annual Cycles. The biosphere also operates on shorter-term cycles such as day/night and the yearly progression of the seasons. Atmospheric composition shows a marked fluctuation in carbon dioxide content between daylight hours when (barring temperate zone winters) plants are actively photosynthesizing, and night when they are only respiring (oxidizing). Similarly, there are wide differences seasonally in air composition reflecting the activity levels of the biosphere. In the Northern Hemisphere between April and September carbon dioxide levels drop by about three percent, reflecting an uptake by photosynthesizers of four billion tons of carbon.

Life on planet Earth has adapted to its rotational circumstances and a broad spectrum of plants and animals show marked circadian (daily) rhythms. Experiments have demonstrated that these frequently continue even when the external stimuli are changed; such as the opening and closing of its flowers even when a plant is kept in continuous light or dark. Bio-ryhthms also govern the flowering of many plants; botanists distinguish between "short-day", "long-day" and "day neutral" plants. The biosphere has also had to accomodate seasonal adaptations to fit the wide range of growing conditions -- from the two month growing season of the arctic tundra to the year-round growth of the humid tropical rain forest; from species triggered to respond to unpredictable desert thunderstorms, to plants whose finely adjusted photoperiod enables them to ignore moisture/temperature signals which may come at an inopportune time for germinating or seeding.

Three major strategies evolved towards yearly cycles are annual, biennial and perennial plants which tend to dominate different environments. Annuals, because of their ability to respond quickly and complete their growth pattern, are favored in the arid regions where having to maintain organic tissues year-round is a liability. In moist warm regions perennials tend to dominate, leaving little space (and light) for the annuals, as is evidenced by the ease with which one can usually walk under the towering trees of temperate and tropical forests.

Diversity. Cybernetics -- the study of control mechanisms in living organisms and mechanical systems -- underscores the importance of diversity to the long-term survival of the biosphere. The Wiener-Shannon-Ashby "necessary diversity law" states that a cybernetic system only possesses stability for blocking internal and external disturbances when it has sufficient internal diversity.[8] To model and appreciate the diversity of the biosphere we can approach it from the standpoint of its taxonomic groupings (the five kingdoms of life), ecological organization (realms, biomes and landscapes), geographic distribution (four films of life), protagonists (species), and the driving force (reproductive pressure).

The Five Kingdoms. Individual organisms range in size from 20 nano-meter (20 billionths of a meter) viruses to

100 meter rainforest trees; in weight, from microbes weighing thousandths of a gram to sperm whales weighing over 100 tons; in life spans, of hours for bacteria (or are they immortal, eternally subdividing from the advent of the "Adam-microbe"?), to bristlecone pines and Japanese cedars which can live over 5,000 years. Generating this diversity modern biology now recognizes five kingdoms of life, based on two fundamental cell types.

Kingdom one, *the Monera,* are the prokaryotic organisms. Lacking a cell nucleus and discrete mitochondria for cell division, the monera are the original life-forms of the planet and ran Biosphere I alone for billions of years.

Kingdoms two through five are composed of eucaryotic-celled organisms. Kingdom two, *the Protoctista* are mainly single-celled microbes. Kingdom three are *the Fungi;* kingdom four, *Animalia;* and kingdom five, *the Plantae.* The monera include bacteria and cyanophytes (blue-green algae). Whereas the metabolism of eucaryotic cells show little variation, that of the monera include chemo- and photosynthesizers, both aerobic and anaerobic, heterotrophs, saprophytes, and most of the microbes able to process mineral elements. The monera were the first life-forms to recolonize after the Krakatoa volcanic eruption of 1883, and after the nuclear bomb testing on Bikini Atoll.

The biospheric functions of the monera are the incorporation of new material in the biotic cycle, regulation of atmospheric gases, and the pioneering of previously barren areas. The monera include the only nitrogen-fixing organisms: soil bacteria and about one hundred species of cyanophytes which fix atmospheric nitrogen outside the soil, on rocks, the bark of trees and even snow.

The protoctista, composed of one-celled eucaryotic organisms and their close descendants, play a role redistributing organic material in water-media and of concentrating particular elements.

The fungi are almost universally saprophytes, essential to the biotic cycle through their decomposition of complex organic macromolecules to simpler forms.

The plant kingdom is composed of red algae, true algae and the higher land plants. Red algae are marine organisms capable of photosynthesis in diminished light which enables them to live at oceanic depths of up to 180 meters. The algae and higher plants are nearly all photo-autotrophs; the rare exceptions being those plants which compensate for soil nitrogen deficiency by trapping and ingesting insects and sometimes small animals; and the partially parasitic plants. The plants' biospheric role is the storehouse of biomass (99 percent) and creation of new organic matter (40 of percent annual biospheric production).

The myriad of animal species -- who number 7/8 of the 1.5 million currently identified -- relates to their metabolic function as heterotrophs. Insect species are the most numerous, followed by molluscs and then vertebrates. Animals' biospheric function relates to their "nomadic lifestyle" -- transporting seeds and fertilizer for great distances, as witness the patterns of migrating birds, fish, dolphins, whales and men. In addition, animals may play a regulatory role by consuming a portion of the production of the autotrophs, which otherwise might escape from the biotic cycle undecomposed. They also contribute by their absorption and excretion -- e.g. filter-feeding marine and soil and sediment detritus-consuming animals. Animal saprophytes accelerate rates of decomposition. As Andrey Lapo puts it, "Losses in the production of the autotrophs is the 'price' the vegetation pays to herbivorous animals for the acquisition of homeostatic reactions ... [for the animals] in the biosphere are its transportation means, control panel, conditioner and sanitation means." [9]

Realms. In the five kingdoms we have identified the principal actors in the biosphere, together with their characteristic roles. Diversity increases yet again when we look at the arenas or stages on which the biosphere performs: realms, biomes, landscapes.

The biosphere has been divided into six biogeographical realms which contain characteristic floral and faunal populations. The Ethiopian, Oriental, Palearctic, Nearctic, Neotropical and Australian realms reflect worldwide climatic zones and the existence of barriers which have tended to impede the distribution of organisms. Some divisions of the realms reflect historical continental locations; North and South America were connected only recently in the evolutionary timescale and belong to two different realms. Within each realm, biota radiate from a population center; on a biospheric level, the Ethiopian (covering most of Africa) and Oriental (Southern Asia and Malaysia) realms have been especially active as radiation centers for the spread of plants and animals worldwide.

Biomes or Biogeocoenoses. Biomes refer to large-scale complexes of life communities, soils and climates in characteristic geographic positions. Biomes can very accurately be predicted in occurence using the variables of latitude and rainfall. The number and names of the biomes vary, but the following are nine major ones: tropical forests, tropical woodlands, tropical savannahs and grasslands, Mediterranean-type vegetation, temperate forests, temperate grasslands, deserts, boreal forests and tundra.[10]

For adaptation to changing Earth conditions, the biomes offer a wealth of competitive alternatives. Some biospheric theorists, like M.M. Kamshilov, view biomes (biogeocoenoses) as having become "the primary unit of evolution", comprising as they do all five kingdoms linked in foodchain biotic circulation, and having the potentiality, if conditions are favorable, of spreading and conquering new areas, but "competing with one another for substance, energy and space they set a limit to expansive tendencies of their rivals ... the steppe strives for replacing the forest, the forest strives for replacing the steppe etc. Naturally only [the] most integrated biogeocoenosis can win the competition, i.e. [those] which exhibit the greatest division of functions between its members and, hence, have a greater number of internal biotic links ... The evolving unit is the specific population which is evolving, however, only as a part of its biogeocoenosis, and the latter, in turn, is an integral part of the biosphere."[11]

Eugene P. Odum, a founder of modern ecological theory, points to this important mechanism in explaining, for example, the persistence of rhododendron or grass areas high in mountain forest regions: "whatever the reason (perhaps fire) for their original establishment, the shrub community is now so well established that it resists invasion by the forest. In this situation we can observe how whole communities, as well as the individuals in them, compete with one another."[12]

This biomic natural selection can be seen in the geologic

record when continental drift changed a region's global position. Thus Greenland's palm tree fossils evidence a time of tropical forest biome; or changing climate -- the Amazon was predominately savannah at a more arid stage in its history. Currently, the desert biome spreads in many regions at the expense of the savannah because overgrazing, unwise agriculture and elimination of tree cover reduces the complexity of the biome which can be supported.

Landscape. As the building-blocks of biomes, physical geographers look at typical "landscapes", or "biogeographical provinces", which are smaller-scale entities characterized by its relief, surface deposits, soils, microclimate, vegetation and animal life.[13] Landscapes are reflected in human consciousness, which they also help to form, as regions, and create the intensest emotional ties. A single phrase: the Kimberleys, the Short Grass Country, the Lake Country, Provence suffices to set off the pleasurable even ecstatic recognition of one of the complex patterns we call a landscape or biogeographical province.

Four Thin Films. The diversity of Biosphere I is also increased by "patchiness", or non-uniform spatial distribution. There are large areas of the planet -- its core deserts, permafrost areas and mid-depth oceanic volumes in which life has scarcely a toe-hold. In the Atacama desert of coastal Peru, the only life in some areas are spores blown in from great distances; there are small communities of monera living hundreds of meters under permanent ice in the coastal waters of Antarctica. These contrast with the hundreds of species of plants, thousands of insects in an acre of rainforest. In biomass production three orders of magnitude separate the barren from the most life-abundant areas. So in fact, the biosphere is not at all a uniform sphere -- but might better be viewed as a "pervasive" on its strata of the Earth.

But the chief means by which the biosphere achieves its non-uniform diversity is by concentrating at boundary interfaces, where ocean meets the bottom, air meets ocean, air meets the soil and in the soil itself. Far from being "a" thin layer of life, Biosphere I consists primarily of four thin films and their interactions.

In the ocean, the factors limiting the biomass are principally the availability of light, and access to a sufficient concentration of nutrients (dissolved salts). Thus life is concentrated in two eco-horizons: first, near the surface, generally to a depth of 100 meters where light is sufficient for photosynthesis; second, the ocean bottom where marine detritus falls and accumulates, permitting a benthic (ocean-bottom) community of chemotrophs, heterotrophs and saprophytes.

These two eco-horizons help explain the importance of shallow coastal waters, marshes and lakes to the biospheric circulation. There smaller separation between surface and bottom, plus additional supply of nutrients from terrestrial ecosystems through erosion, river deposition etc. leads to an exceptional abundance of life.

On the land, life is concentrated, first, on and near the surface, and second, in subterranean soil levels. The recent evolution of woody trees and birds marks a conquest of near air space for the biosphere, but the surface eco-horizon rarely exceeds a height 200 to 300 feet above the ground. The aerobic and anaerobic soil life greatly exceeds the surface: one gram of forest soil contains 400 million bacteria, 2 million fungi, 100 thousand microscopic algae and 10 thousand protozoa. Important to soil life is the penetration of air facilitated by plant roots, earthworms and burrowing animals. Soil air composition is determined by the intense activity of the microbes. Below the soil level which averages 1.5 meters (five feet), the only life are the anaerobic bacteria, and although they have been discovered recently as deep as 10,000 feet, life is concentrated near the surface.

These four thin films of life play varying roles in the biosphere. The phytoplantkon has been called the "kitchen garden" of the ocean, renewing itself *daily*, producing the microscopic food which sustains all marine life. The benthic is the "storehouse of its finished products"; its high pressure, oxygen-poor conditions being much more suited for deposition of organic matter than terrestrial eco-horizons. This deposition is even more accentuated in swampy and shallow coastal areas where bacterial production of methane ensures the absence of oxidizing conditions necessary for decomposition and where detritus can quickly reach the bottom to be safely buried. This marine sedimentation due to the evolution of calcium-shelled marine organisms some 550 million years ago produces the vast deposits of chalk and limestone; and secondarily, from diatoms, deposits of silicates.

The land is home to the woody plants (where nearly all the biomass of Biosphere I is concentrated) and its dependent animal population who serve as dispersal agents for this biospheric wealth. The soil eco-horizon works to bring new inorganic material into the biotic circulation and plays a big role through its dense microbial population in atmospheric and material cycle regulation. Atmospheric composition seems to be in equilibrium with the biosphere's soil air.

The first spectacular photographs from space of the "Blue Planet" underlined what Arthur Clarke proposed: that the third planet from the Sun should rightly be named "Planet Ocean". The coral reefs, which occupy about 600 thousand square kilometers, were known to be highly productive, but were still underestimated by an order of magnitude until the work of Walter Adey at the Smithsonian Marine Laboratory during the past decade. Oceanographers now see that they function as a major regulator of ocean nutrient levels by bio-filtering the entire volume of the ocean on a cycle of 40,000 years. Similarly, the location of nutrient-upwelling areas has only recently been systematically mapped and their importance ascertained. For example: the Peruvian coastal upwelling with an area of 0.2 percent of the ocean produces some 15 to 20 percent of the annual world fish harvest.

Species. Biosphere I entered a new era of species proliferation when multi-cellular organisms, having arisen on their eucaryotic base, and with their microbial symbionts, successfully spread onto the land around 400 million years ago. This exploitation of opportunity reflects in land species representing over 90 percent of the biosphere's total, due to land's greater range of econiches. Also characteristic of terrestrial ecosystems is the phenomenon of succession: the sequence of developmental stages of plant/animal associations which becomes more complex, and tends towards the eco-climax, the greatest species diversity and biomass productivity possible under the given circumstances.

Though the fossil record of evolution is only relatively

detailed for the past 500 to 600 million years, its broadstroke direction reveals interesting regularities. Each species' evolutionary opportunity relates to its biospheric-realm-biomic-landscape-ecosystem-econiche position and its genotypic and phenotypic ability to participate more competitively and efficently in the biotic circulation. Wide variations exist in the evolutionary potential of genera, "speciose" genera giving rise to many closely related types, and "depauperate" genera, relatively few. There also appears to be a life-cycle of genera, with full vigor of radiative branching occuring during the first quarter of its life, followed by a static period with little or no change, and finally through succession by one of its offspring lines (pseudoextinction) or elimination by competing lineages for its place in the biotic cycle (true extinction), it comes to an end.[14] Or as Kamshilov elegantly writes the epitaph: "Biosphere has cancelled them from the list of beings worthy of its attention".

The stability of ecosystem, landscape and biome, and thus the higher integration, Biosphere I, is enhanced by the millions of current and oncoming species because many genera have their "candidate species" ready to occupy and compete for their segment in the web of life. Should mass extinction occur, genetic potentialities have been available and probably are still available to evolve new morphological forms to refill the vacated econiches.

The prokaryotes deserve a separate note in the question of species diversity, for it is now known that they possess the ability to transfer genetic material; or as Clair Folsome wryly comments, there may be "as many types of microbes as there are microbes". Current biologic convention is that microbe species are classified by their physiology and metabolism, in pure laboratory culture, but, in an important sense, they do not exist in "nature". Lynn Margulis and Dorion Sagan comment on microbial power for genetic exchange and adaptation: "as a result of this ability, all the world's bacteria essentially have access to a single gene pool and hence to the adaptive mechanisms of the entire bacterial kingdom. The speed of recombination over that of mutation is superior: it could take eucaryotic organisms a million years to adapt to a change on a worldwide scale that bacteria can accomodate in a few years ... The result is a planet made fertile and inhabitable for larger forms of life by a communicating and cooperating world-wide superorganism of bacteria." [15]

Evolutionary trends continue, a drama that almost surely has not reached its climax of environmental change, genetic mutation, and phenotypical fitness.

Some botanists have discerned the beginning of a new trend in plant evolution away from co-evolution with insect pollinators which has characterized the history of flowering plants. Now there is mounting evidence of the spread of asexual and apomictic (without fertilization) reproduction, reducing plants' dependence on insects, weather conditions etc. This is being observed in varied genera and geographical locations indicating a "new stage in biosphere evolution characterized by a larger autonomy in the development of the species of flowering plants".[16]

Species diversity is the rainbow on time's palette with which the biosphere paints.

The Life Force. "The pressure of life" was the term Vernadsky used to describe the fundamental driving force in biospheric evolution, the relentless urge of life to reproduce itself. "Life has spread over the Earth's surface as a result of its pressure on the surrounding medium. This spreading has been accompanied by adaptation, whereby organisms can accustom themselves to conditions which would have been fatal to previous generations. The limits of the biosphere are therefore not immutable or permanent."[17] Perhaps the historical meaning of Neil Armstrong's first descent to the lunar surface is "One small step for me, one giant step for the biosphere."

We see this life force at work in the organism's production of many, sometimes hundreds or even thousands of seed, spore or young, of which at equilibrium population only one will survive because of competition or lack of resources. A microbe reproducing hourly would produce about eight million progeny in a day if all survived. We see it also in the waxing and waning of biomes which give the biosphere "muscle tone". Modern biological thought gives increasing appreciation to the co-evolution of species, even predator/prey and plant/herbivore relations which serve as evolutionary spurs to both parties. Similarly, depending on their evolutionary history, biomes and landscapes can be less or more complex and developed. "There seems to be a strong relationship between the complexity of a [biome] and its ability to withstand diverse external effects".[18] So, too, for Biosphere I as a diverse synergy of varied realms, landscapes, and biomes, kingdoms, and species.

As Charles Darwin formulated it, "A struggle for existence inevitably follows from the high rate at which all organic beings tend to increase ... Hence, as more individuals are produced than can possibly survive, there must in every case be a struggle for existence, either one individual with another of the same species, or with the individuals of distinct species, or with the physical conditions of life."[19]

Functions

> After the scale and history of a given phenomenon is grasped which gives a handle on its reality, the next necessary idea is its function in the universal process.

Free Energy. The biosphere performs a unique and crucial function on the planet: it acts as an apparatus for generating and storing free energy, the thermodynamic measure of the energy available to do work, with which it organizes, upgrades and transforms itself and its environment. The inanimate world on planet Earth would, of course, remain an equilibrosphere without the biosphere, like Mars or Venus; its state, material and chemical composition predictable by physical science equations, that is to say, a state characterized by the minimum of free energy under the given conditions.

John von Neumann, the mathematical theorist, in analyzing organic and mechanical self-reproducing systems discerned the principle that "complication, as well as organization, below a certain minimum level is degenerative and beyond that level can become self-supporting and even increasing." This organized complexity is the hallmark of Biosphere I and its increase is the result and further cause of the driving force of evolution. Some life scientists follow the lead of British biologist J.D. Bernal and view the biotic circulation itself, the close coupling of the metabolic and decomposing processes, as the origin of life systems on the planet, not the rise of individual organisms. This led Kamshilov to the thought-provoking concept: "Life came into being before living organisms did".[20]

In any event, from its beginning the biosphere has served literally as a solar energy collector, an apparatus for generating the free energy that it applies to evolving not only itself but to reshaping the planet's equilibrosphere and eventually to give a headstart to Homo sapiens' creation of a technosphere.

We can read the history of this application of free energy in evolving greater complexity in the change from the earliest aeon of Biosphere I -- the fermenting phase of chemo-synthesizers who worked with the initial wealth of carbon and hydrogen compounds. When that supply became scarce, photosynthesis developed, first anaerobically using hydrogen gas and hydrogen sulfide (H_2S). Next, evolved the capacity to split photosynthetically the abundant water molecule, releasing oxygen as a dangerous by-product. Then further life-forms evolved which could work with oxygen for cellular respiration. The eucaryotes with their co-evolved mitochondria, which began as symbiotic prokaryotes before their assimilation inside the cell, were the first "internal combustion" engines on the planet -- and achieved an increase in energy production over ten times that of the early fermenting organisms. The relentless "pressure of life" too is a way that the biosphere gains access to new sources of energy to upgrade, finding new territories, new chemical capabilities, new raw materials, in short new econiches to elaborate ever more complex and organized communities.

Schrodinger, like Buckminster Fuller, saw in the biosphere a powerful force combatting entropy: "To live is to suck orderliness from the environment." The open system we call Biosphere I uses its input of solar energy to create order in the material cycling of the planet -- incorporating new inorganic material in its processes, concentrating and depositing minerals, and maintaining a non-equilibrium atmosphere. The two to eight percent of its annual product which is not immediately recycled into living forms but is deposited in the Earth's crust or added to its atmosphere amounts to some 6 to 16 billion tons of material annually, which over the aeons create the "megabiosphere": which encompasses about the top 20 miles of the Earth's crust to the top of the troposphere.

Without life, as Vernadsky saw, "known minerals would irrevocably disappear. There then would be no force on the Earth's crust capable of perpetually giving birth to new compounds. A stable chemical equilibrium, a chemical calm, would be established ... [but] life exerts a powerful permanent and continuous disturbing effect on the chemical stability of our planet."[21] The chemical fecundity of life dwarfs that of inorganic processes -- there are an estimated two million organic compounds in Biosphere I as compared with 20 thousand inorganic substances. Richard Evans Schultes, the intrepid Amazonian ethno-botanist of Harvard's Botanical Museum, has pointed out that the tropical rainforests of the biosphere serve as its pharmacopia. On average, ten percent contain chemicals with active metabolic properties, and Schultes warns that if its great species diversity is lost, we will also lose potential medicinal and industrial resources of inestimable value.

The biosphere developed its chemical properties in ways which far surpass those of the latest laboratories. The evolution of life's catalytic enzymes enables organic reactions to proceed thousands, even millions, of times faster inside the living cell than does an uncatalyzed reaction. Life also evolved methods of performing its metabolic work under much less extreme conditions -- in living cells the oxidation of carbohydrates and fats is done at less than 100 degrees F (37 degrees C) whereas in the laboratory temperatures of over 1000 degrees F (450 - 500 degrees C) are required. The synthesis of ammonia from gaseous nitrogen is routinely accomplished by bacteria at normal temperatures and pressures while to accomplish it industrially requires not only a temperature of 500 degrees C but pressures 300 times greater than that at sea level.[22]

The organizing function of Biosphere I can be seen most dramatically in its billions-year evolution towards greater complexity of individual species, biotic circulation, and ecosystems, all of which require an increasing ability to handle and transmit information. Charles Darwin wrote: "... the structure of every organic being is related, in the most essential yet often hidden manner, to that of all other organic beings, with which it comes into competition for food or residence, or from which it has to escape, or on which it preys".[23]

This information network, for example, links the defense mechanisms of prey or plant with offensive innovations of hunter or herbivore so that they are both pruned of weaker members, and forced to match the other's advances. Biologists have ironically called these relationships the world's first "arms races" but the effect is a dynamic improvement in biospheric resiliency. Natural selection ensures the preservation of those species which contain the information-coding that most favors survival of its progeny.

This increase in information can be seen in the biosphere's evolution of muti-cellular, multi-organed lifeforms. From the single-celled organism to Homo sapien sapiens, with his 110 quadrillion cells. Characteristic, too, of the increasing complexity of operating in the biosphere, one hundred quadrillion of those "human" cells are those of microbial symbionts who perform essential roles in digestion, excretion, metabolism and without whom we could not survive. Ninety-five percent of modern land plant species are co-evolved with fungal root partners, the mycorrhiza, essential for their nutrition.[24] "The faculty for absorption of information is a means for enhancing self-regulation ... Then organisms which absorb more information must possess certain advantages. Hence, natural selection must promote store of information, that is, growth of organization's complexity".[25]

As the cycling of material is the chief activity of the biosphere, evolution tends to favor the more developed and stable ecosystems, those which succeed in a more productive utilization of substance, energy and information. The ash content of plants shows a steady increase, reflecting the selective advantage of those species better able to extract and utilize inorganic soil minerals. The energy fixation of the multi-canopied forest, where competition for light is intense, greatly exceeds that of other biomes, and barring disturbance, forest biomes will outcompete others.

Plants have also developed two energetic pathways in

photosynthesis, that of the C_3 and C_4 plants. The C_4 plants are better able to work with high intensities of light and temperature, and require half the quantity of water that C_3 plants demand. C_4 plants dominate in tropical grasslands and desert regions, but perhaps surprisingly in view of their physiological superiority, they are outcompeted in regions of mixed species and average climatic conditions, and C_3 plants make up most of the biosphere's plant population. As Odum notes, natural selection favors the whole matrix which is more advanced, not necessarily the "genius", outstanding species; "survival of the fittest in the real world does not always go to the species that is physiologically superior in monoculture."[26]

Life-Enhancing Feedback System. The biosphere also functions as a cybernetic system moderating environmental conditions on our planet for the continuance and spread of life. We can see this in operation in the re-colonization of volcanic or degraded areas. As succession follows succession, harsh initial conditions of lack of soil, lack of shade cover, rudimentary food chains give way to more and more diversified biota, able to increase, and by their increase creating more econiches. Life begets and sustains life.

Planetary climate mechanisms hold great interest for students of the biosphere. Although there are great unknowns and considerable impact from cosmic and geologic forces, that the Earth's average climate has varied only within a relatively few degrees, and has certainly avoided the extremes of freezing or boiling since the advent of life, points to vast homeostatic mechanisms within the biosphere. There are similar factors at work in the major gas and mineral cycles where the biosphere recycles and maintains the chemical disequilibria which favors life's flourishing.

For example, carbon dioxide would be injurious to life if present in quantities exceeding one to two percent and furthermore, as one of the principal "greenhouse" gases of the atmosphere, its increase leads to a rise in temperature. But these higher levels of carbon dioxide also enable plants to increase their productivity thus taking more carbon dioxide out of the atmosphere to be built into organic compounds, producing greater deposition of organic matter, taking yet more carbon out of circulation. Whether this biospheric cycle can respond adequately to the relatively sudden carbon dioxide buildup produced by man's industrial activity is a question.

Though the Earth has endured at least three major periods of Ice Ages, where worldwide temperatures declined and glaciation covered much of the land mass of the planet; some stabilizing forces introduced negative feedback and prevented the cooling from deepening and finally destroying the biosphere.

Similarly, though solar output is estimated to have increased by at least 30 to 50 percent (some scientists maintain it may have more than doubled) since the origin of the Sun 4.5 to 5 billion years ago, the biosphere has undoubtedly acted to help regulate planetary temperature conditions within a small oscillation.

Inevitable Doom of Biosphere I

If we take a long-range view of the future course of Biosphere I it becomes clear that if it cannot transcend the conditions of being confined to planet Earth, it will come to an end long before the end of the universe and thus will have attained only the historic status of a local blind alley. When that will be depends on which of the following factors goes critical first.

Sun Changes. If current astrophysical theory is correct, we may expect a total life-time for a star with a mass like our Sun of approximately ten billion years. About five billion years from now our Sun will swell into a red giant as the nuclear materials in its core begin to be exhausted. The Sun will then expand, filling most of the volume of the inner solar system, consuming Mercury and Venus, and perhaps the Earth. Even if it does not, the radiation from our swollen Sun will boil our oceans, raise surface temperatures to 4,000 degrees F and finally evaporate its core. Further processes within the Sun, the collapse of its core when nuclear materials diminish further, will lead to a second stage of helium fusion which will then power the Sun for another 100 million years. And when helium is gone, the Sun will collapse to the white dwarf stage and then burn out. There is another end for stars, not thought likely for our Sun because of its medium mass, to explode as a super-nova, destroying all around it.

This future of five billion years is probably the best-case scenario; it may well be that increased radiation from the Sun will exceed the biosphere's ability to moderate much sooner than that. Some scientists have suggested that in about 300 million years, biospheric regulation will prove unable to cope with the increased heat load.

Cosmic Impacts. The early history of our solar system was marked by catastrophic impacts by the "left-overs" from its creation: planetesimals, comets and meteors. They probably contributed significantly to the availability of carbon-rich compounds which fueled the ancient biosphere. With time, these impacts became much more infrequent; but not unknown.

On June 30, 1908 a giant fireball was observed in central Siberia followed by an enormous explosion. Two thousand square kilometers of forest trees were leveled. This became known as the Tunguska Event and was probably caused by part of a comet hitting the Earth, which astronomers estimate was 100 meters wide, weighed one million tons, and traveled at 70,000 miles per hour.[27] Our solar system is still littered with such objects; one trillion comet nuclei are orbiting the Sun in the vast Oort cloud beyond Pluto. Depending on Earth's orbit and the movement of our solar system within the Galaxy, periodically some of these may be gravitationally unleashed from their present orbits and sent on a voyage towards the Sun and the inner planets. There are also the increasing numbers of observed Apollo asteroids, possessing orbits which cross that of Earth. A "small" impact like the Tunguska probably occurs on Earth once every thousand years, but an impact with a comet the size of Halley's might occur every billion years.[28]

The implications of these cosmic impacts on the evolution of our biosphere is still unknown though many scientists are advancing evidence that such an event may have triggered the Triassic extinction of the large reptiles --

an unusually high incidence of radioactive iridium, rare on Earth, but common in meteors has been found dating from that time. Other theories advanced look to the effects of a "nearby" (within 10 or 20 light years) supernova explosion which would have "sprayed an intense flux of cosmic rays into space ... [which] entering Earth's envelope of air would have burned the atmospheric nitrogen ... removing the protective layer of ozone ... frying and mutating the many organisms imperfectly protected against intense ultraviolet light".[29] The size of cosmic impact which would so disturb conditions as to lead to total extinction is unknown -- and its probable occurence at present incalculable. But the danger exists.

Technics Abused. The possible scenarios in which man's abuse of technics might lead to the death of the biosphere have aroused great interest since the international conferences on the aftermath of a full-scale nuclear war. "Nuclear winter", if it occured in its most extreme modeled form, would certainly be the single greatest disaster ever to befall Biosphere I, and could lead to an extinction of virtually all higher plants and animals, terrestrial and marine. It could set the evolutionary process back millions of years and perhaps dead-end it in a glorious microbial sunset pre-technospheric level, absent a big-brained, manually agile, complex-culture creating being. Still it almost certainly would not eliminate the monera and protoctista, some of which are resistant to radioactivity, and whose numbers include enough chemotrophs and saprophytes who would survive even years of darkened skies; the fungi, capable of remaining viable in spore form; and even many insects. Of course, our hopes, dreams, and possibilities would no longer possess even nostalgia.

The human-driven technosphere has paralleled the sun-driven biosphere from which it arose and has become a second major force of transformation on the planet. By 1970 over 20 tons of raw material were extracted for every human. Although a product of the biosphere, biologically mankind has manifested as a "plague species"[30], unable thus far to limit exponential population growth and equilibrate industrial activities and waste-products. The destruction of the rainforests of the new world has already eliminated many species and, if unchecked by the turn of the century, could become the single greatest mass species-extinction event. The toll of man's agriculture, only ten thousand years old but widely recognized as the most serious planetary desertification factor, continues at rates of annual soil loss of 150 tons per square mile in Asia, 245 tons in North America, 160 in South America and 90 in Europe.[31]

Vernadsky discerned a new incipient phase in biospheric evolution -- the noosphere, or sphere of intelligence, wherein humanity could employ its evolutionary gifts as a creative collaborative agent of evolution -- and where the widening conflict between technosphere and biosphere could be transformed into a synergy. This still remains possible, and in certain fields has already begun, as we shall see in later chapters. But for the moment we focus on the threats caused by man's current unintelligent use of biospheric resources and technics.

The consequences of the pollution emanating from industry is probably the most serious threat to go beyond impoverishing to imperiling the biosphere. Concern centers now on depletion of the ozone layer by industrially-generated fluorocarbons and nitrogen oxide; acid rain from sulphur dioxide pollution; and a rapid increase of carbon dioxide from fuel combustion in recent decades. All of these pose dangers that the resilience of the biosphere may prove unable to compensate for such rapid, in evolutionary time-frame, disturbances, and that fundamental safeguards, such as the ozone layer, the mix of reactive gases in the atmosphere, the biologic detoxification of dangerous substances (heavy metals, radioactive materials, pesticides, etc.) which concentrate in the food chain, will fail with the same system simplifying results for the entire biosphere as "nuclear winter".

Failure of Homeostasis. In addition to man's recent and increasing impact on the grand cycles and mechanisms of the biosphere, too little is known of the natural feedback loops at work to be able to anticipate the kind of ultimate stress Biosphere I could survive -- though it has passed every previous test. "If stress is severe enough, only tolerant organisms survive ... Darwinian natural selection is the ultimate ancient Gaian feedback system upon which all the newer technological and behavioral ones are based ... planet-wide living systems of temperature and atmospheric regulation can only be guessed at ... [but] from a planetary perspective they are robust".[32] Nonetheless, no system can be completely fail-safe from internal run-away processes, nor completely catastrophe-proof against externally caused catastrophes.

Combination of the Above. The final threat to the evolving future of the biosphere comes from the combined effects of several of the above threats -- where the health of the system, weakened by one factor, perhaps a severe cosmic impact, or disruption of a major cycle, proves incapable of integrating yet another.

The future of Biosphere I may then, potentially, be measured in the billions of years its solar energy source will last; or be truncated by compound disaster on a cosmic, geologic and human level; but in either case its lifetime will be confined and limited to that of planet Earth. It is doomed to die unless it can birth offspring that can escape to other stars.

REFERENCES

[1] Vernadsky, Vladimir, *The Biosphere*, Synergetic Press, London, 1986, p. 17.

[2] Margulis, Lynn and Karlene U. Schwartz, *The Five Kingdoms of Life*, W.H. Freeman, San Francisco, 1981, p. 26.

[3] Milne, D., Raup, D., et.al, *The Evolution of Complex and Higher Organisms*, NASA SP-478, Washington, D.C., 1985, pp. 14-15.

[4] Lovelock, James, *Gaia: A New Look at Life on Earth*, Oxford Univ. Press, Oxford, 1979, pp. 68, 74-79; and C.C. Delwich, "The Nitrogen Cycle", *The Biosphere*, Scientific American, W.H. Freeman, San Francisco, 1970, pp. 75-76.

[5] Lapo, A.V., *Traces of Bygone Biospheres*, Mir Publishers, Moscow, 1982, p.93.

[6] Ibid., p.99.

[7] Odum, Eugene P., *Ecology*, 2nd ed., Holt, Rinehart, and Winston, London, 1975, p. 104.

[8] Lapo, *Traces of Bygone Biospheres*, p. 25.

[9] Ibid., pp. 74-75.

[10] Myers, Norman, "Biomes", *The Biosphere Catalogue*, Synergetic Press, London/Tucson, 1985, pp. 9-13.

[11] Kamshilov, M.M., *The Evolution of the Biosphere*, Mir Publishers, Moscow, 1976, pp. 95-96.

[12] Odum, *Ecology*, p. 199.

[13] *Soviet Geography Today: Aspects of Theory*, Article by K. Markov, Progress Publishers, Moscow, 1981, p. 89.

[14] Milne, et.al., *Evolution of Complex and Higher Organisms*, p. 12, 13.

[15] Margulis, Lynn and Sagan, Dorion, *Microcosmos*, Summit Books, New York, 1986, p. 17.

[16] Kamshilov, *Evolution of the Biosphere*, pp. 214-215.

[17] Vernadsky, *The Biosphere*, p. 55.
[18] Kamshilov, *The Evolution of the Biosphere*, p. 91.
[19] Darwin, Charles, *The Origin of Species*, 6th Ed. Oxford Univ. Press, London, 1872, p. 77.
[20] Kamshilov, *Evolution of the Biosphere*, p. 39.
[21] Vernadsky, *The Biosphere*, p. 17.
[22] Lapo, *Traces of Bygone Biospheres*, p. 20.
[23] Darwin, Charles, *The Origin of Species*, p. 77.
[24] Margulis, Sagan, *Microcosmos*, pp. 67, 171-2.
[25] Kamshilov, *Evolution of the Biosphere*, p. 226.
[26] Odum, *Ecology*, p. 78.
[27] Sagan, Carl, *Cosmos*, Random House, New York, 1984, p. 73-76.
[28] Ibid., p. 82.
[29] Ibid., p. 283.
[30] Sharma, Roland, "Jungle Wildlife" in *Man, Jungles and Survival*, Synergetic Press, London, 1980, p. 18.
[31] Odum, *Ecology*, p. 179.
[32] Margulis, Sagan, *Microcosmos*, p. 274.

2

Biosphere II

Evolution to Biosphere II

To give historical perspective to a way that Homo sapiens and Biosphere I can transform themselves from localized planetary lifespans to cosmic immortality, we will review the main stages in cosmic evolutionary history to see where our biosphere came from, and where we could be headed.

Big Bang. At a time perhaps 18 billion years ago, all our universe was concentrated in a very small and very hot (over 100 billion degrees K) point. Physicists have given theoretical figures in the nanoseconds after the Big Bang explosion which ensued -- at 10^{-23} seconds, temperatures were greater than 10^{30} degrees K, and concentrations were 10^{50} grams/cm^3. These almost unimaginable conditions precluded the existence of matter in any form, until after 10^{-20} seconds when rapid cooling permitted heavier elementary particles, protons, neutrons and mesons, to form. At close to one second after the Big Bang, temperatures dropped to 10^{10} degrees K (10 billion degrees K), and concentrations to 10 billion grams/cm^3. Now lighter particles, electrons, neutrinos and muons coalesced to self-annihilate into photons fueling the radiative fireball.

This first phase of cosmic history, which lasted about 100 seconds is known as the Radiation Era, since the matter present was of minor significance compared to the exploding core of dazzling light.

Next followed the beginning of the Matter Era whose first stage was the atom epoch, atoms being the first stable form of matter that could exist until about one million years after the Big Bang. During this period, radiation began to be eclipsed by matter as temperatures and densities continued to fall, reaching 10^6 degrees K (similar to temperatures in present-day stars) and 10^{-10} grams/cm^3. With the formation of electromagnetically neutral atoms, much more resistant to radiation, "matter had, in a sense, over-thrown the cosmic fireball."[1]

Hydrogen, the simplest and first of atoms, formed in abundance and still is predominant in the universe. Smaller quantities of helium were also formed from the fusion of two hydrogen atoms until temperatures fell below 10^7 degrees K. Heavier atoms could not form since, by the time there were enough helium atoms present, the cooling cosmic conditions precluded their fusing.

Formation of Galaxies and Stars. The gravitational coalescence of these enormous clouds of hydrogen/helium created the galaxies, in the era from one million to perhaps ten billion years after the Big Bang. Astronomical observation indicates there has been no galaxy formation after this time. Average conditions in the expanding universe by the middle of the galaxy epoch reached 3,000 degrees K, 10^{-20} grams/cm^3 . Gravity and the law of the conservation of angular momentum combined to shape the galaxies into their frequently flattened, spiral-arm forms. An estimated 100 billion galaxies formed during this period.

Within these coalescing "lumps of matter", smaller concentrations cohered gravitationally, and as internal temperatures rose, the first stars ignited. This process, which began shortly after the first galaxy formation, is a continuing one: modern astronomers have observed the creation of new stars within galaxies.

The atom, galaxy and stellar epochs constitute the three steps of the second phase of cosmic evolution: the Matter Era, as our universe became ever darker, cooler and thinner.

First Generation of Stars. As the first clouds of matter gravitationally contracted, they reached critical density and sufficient heat to start hydrogen fusion -- the first stars "turned on" -- producing helium nuclei and additional heat and pressure which (as long as the supply of the star's hydrogen lasts) prevents further collapse.

The production of all elements heavier than hydrogen/ helium can only occur just before the death of stars when further implosion generates more than the ten million degrees K required for hydrogen fusion. The lifespan and death of a star are determined by its mass -- the heavier the star, the brighter it burns, and the more violent its end. A star of 20 solar masses (our Sun = 1 solar mass) will be ten thousand times as bright as our Sun but can expect a lifetime of only 20 million years.[2] Stars with smaller solar masses than the Sun will enjoy correspondingly longer but cooler and duller lives; one with a solar mass of one-tenth theoretically shines for a trillion years.[3]

The collapse into smaller cores increases star temperatures to a hundred million degrees K permitting helium nuclei to fuse, producing carbon and oxygen nuclei. For stars of moderate mass this "helium flash" phase is their final burst of energy, after which they cool to white dwarf and finally black dwarf cinders.

But for more massive stars, there is enough material left in the core to continue the collapse/heat increase phases. The novae stars have enough material to fuse carbon nuclei, producing matter up to the very stable iron nucleus. But stars with sufficient solar mass end their lives in spectacular supernovae explosions triggered by densities that fuse protons and electrons into neutrons. The ensuing final collapse and explosion briefly outshines an entire galaxy of hundreds of billions of stars, is powered by enough energy to make small amounts of all the elements heavier than iron, and blast the stellar material over vast areas. About one supernova explodes each century in the Milky Way galaxy.

Second and Later Generation Stars. All of the later generation stars of the stellar epoch have had access to these heavier elements, which though they make up less than one per cent of the matter in the universe (hydrogen is 92 percent and helium 7+ percent) are necessary for planets and life. Our Sun is probably a third generation star, judging from the amounts of radioactive materials and heavier elements in its planetary system. In addition, evidence exists that a nearby supernova exploded shortly before the birth of our Sun. Indeed, Carl Sagan notes, "this is unlikely to be a mere coincidence; more likely the shock wave produced by the supernova compressed interstellar gas and dust and triggered the condensation of the solar system."[4]

Formation of Planets and Moon Systems. Stars coalesced; planetary systems formed. If this happened for even one star in a thousand there would be one hundred million planetary systems in our galaxy alone.

We can look at our solar system to discover underlying dynamics and processes which generally govern planetary systems as modified by each particular configuration of sizes and orbital distances.

Planets nearer their star possess scant water and lighter gases and grade into the outer planets composed principally of such substances.

The planets in our solar system can be divided into the smaller, "terrestrial" inner ones; and the giant gas giants of the outer solar system. The terrestrial ones (Mercury, Venus, Earth and Mars) contain only thin atmospheres as their gravity was unable to hold the initially abundant hydrogen/helium under the impact of the greater heat and radiation of the Sun. The outer planets had gravity sufficient to hold much of their lighter gases against the distant solar radiation; with small, rocky cores, and huge atmospheres, their overall chemical composition more closely resembles that found throughout the universe.

The establishment of a relatively stable equilibrosphere (solids-liquids-gases) occurs during the first few hundred million years of the planet's life. Before that time, the planets are in flux under the combined impacts of the exceptional radiation of a newly-formed star (which has not yet settled down to its "main sequence" rate of fusion), the cooling of the planet's parent materials, and the intense bombardment of asteroids and other "debris" left over from initial planetary formation. Heat generated by gravitational pressure and radioactive elements in planets like Earth powers the shifting of mantle plates and consequent continental drift -- phenomena not true of smaller planets like Mars with a cooler core.

Earth's relatively small gravity, combined with the intense radiation of our young Sun allowed most of the lighter gases to escape our atmosphere. The molten interior of the planet later produced massive outgassing from volcanoes and cooling rocks and a second atmosphere was formed, probably dominated by carbon dioxide, ammonia and water. When Earth cooled sufficiently, the first rains of the planet fell, and oceans filled, where complex organic molecules could be built and circulated.

Origin of Biospheres. Life arose on our planet at least 3.5 billion years ago, one billion years after planetary formation and perhaps 500 million years after its surface cooled. Earthlife may not be late-comers and could well be among the first-starters in the universe; planets of first generation stars could only have been gas giants lacking heavier elements, and massive stars with short lifetimes lack enough time on the relatively stable main sequence to nurture life on its planets. It may only be on medium-sized third and later generation stars that multicellular life could arise and develop.

The Earth's biosphere trends towards increased productivity, diversity, complexity, beauty, and evolutionary potential within the limitations of changing planetary and cosmic boundary conditions.

The third great cosmic period, the Life Era, begins in the Space Age to show the potentiality to transform and manipulate matter, on a universal scale, just as the Matter Era succeeded and dominated the earliest Radiation Era. Whether this will actually occur stands for us as the great drama of the next universal epoch, though some thinkers speculate that this drama may have progressed further in other solar systems.

Technosphere. Within the forty thousand years of modern Homo sapien sapiens, human culture with its power to impact and change the environment has unleashed a new vector of increasing strength on the planet: the technosphere. Baboons and chimpanzees are observed in the wild fashioning and utilizing simple grass and stick tools, and doubtlessly human technics began with such extensions of the human organism; later modifications, such as shelter and clothing, altered the inner and outer environment, enabling man to thrive beyond the range of his biological physiology. The speed of technical innovation compared to evolutionary adaptation has led the technosphere to rival the biosphere in the movement of mass upon the planet Earth within two hundred years of scientifically based expansion.

Techne, originally one of the Greek gods (etymologically: to fashion, make) has grown by exponential progression as mankind spread from its tropical savannah birthplace to rival the monera in mastery of the econiches of the planet. The technosphere now forces ecology to deal with two new and spreading biomes: the agricultural landscape, and the urban connurbation. Ecotechnics (the technics of ecology and the ecology of technics) is born.

Lewis Mumford in his classic, *Technics and Civilization,* points out that the "Industrial Revolution" was merely one phase in a long progression of technological advance: "all the critical instruments of modern technology -- the clock, the printing press, the watermill, the magnetic compass, the loom, the lathe, gunpowder, paper, to say nothing of mathematics, chemistry and mechanics -- existed in other cultures." Man's development of technics also relied from the beginning on utensils (baskets, pottery), apparatus (dying vat, kiln) and utility (reservoir, aquaduct, roads and buildings) as well as "machines".[5]

The technosphere, although primarily sustained by the biosphere, the primary planetary wealth and free-energy producer, has now also learned to extract energy directly from the Sun to power its growth. The Industrial Revolution climaxed the advent of what Mumford termed a "machine culture"; where the technosphere took a rival center of evolutionary significance, "the displacement of the living and organic took place rapidly ... for the machine was the counterfeit of nature; nature analyzed, regulated, narrowed, controlled by the mind of men. The ultimate goal of its development was ... not the mere conquest of nature but her resynthesis: dismembered by thought, nature was put together again in new combinations: material syntheses in chemistry, mechanical syntheses in engineering. The unwillingness to accept the natural environment as a fixed and final condition of man's existence ... from the 17th century this attitude became compulsive."

To invent became a duty, to use ever more modern technics was definitely the way to pay, profit, power, and prestige, if not to progress. The technosphere gained a life and drive of its own -- and increasingly began to come into conflict with that of the biosphere. The macabre apotheosis of this dichotomy can be seen in the nuclear weapon arsenals, twenty tons equivalent of TNT for every human organism, and the invention of the neutron bomb which destroys life while leaving buildings and machinery intact, doubtlessly a subconscious wish-fantasy of more than one of the less psychologically balanced lords of the technosphere.

The technosphere has passed through three major eras: the eotechnic, the earliest period where wood was the characteristic material employed, and water-power its chief energy source; the paleotechnic where iron and coal dominated; and early in the twentieth century marked the transition to neotechnics, which featured use of alloys and electricity. Mumford predicted the fourth step would be "biotechnics" -- technology based on the laws of life, made possible by the ephemeralization of modern technics, doing more with less, which has indeed begun with the computer, based on a model of the brain, and the consumer or post-industrial society based on models of stimulus-response and ego, id and super-ego. The Institute of Ecotechnics sees the next era as that of noogenics, the making of intelligence capable of synergizing the biosphere and technosphere.

Noosphere. The idea of the noosphere was first developed by Vernadsky in the 1920's, then teaching at the Sorbonne. A philosophically idealistic, non-scientific version was promulgated by one of his students, Teilhard de Chardin.

Vernadsky developed the concept scientifically, seeing the noosphere as an integral attribute and new stage in the biosphere's evolution, the stage of intelligent adaptation of relations between man and nature.

The incipient noosphere can function as a reconciling factor in the conflict of technosphere and biosphere. The dangers posed by human population explosion, deforestation, desertification, industrial pollution, degradation of natural biomes and extinction of species all point to the central issue facing the noosphere: whether man can take the biologically unprecedented step of intelligently regulating the technosphere and species expansion, harmonizing it within the biospheric totality, rather than having the biosphere control this species by its well-known "boom and bust" mechanisms. To accomplish this requires selection of the requisite memes (the unit of cultural behavior as genes are of physiological behavior) and their introduction on a mass scale throughout the cultures of the biosphere. For example, memes of the pattern called "naturalist" to replace the memes of the pattern miscalled "conquest of nature", because, of course, it is impossible for the part to conquer the whole. This "mutation of memes" would in no way harm the higher themes of any culture.

Teleosphere. The teleosphere can be defined as the realm of goals and directions of cosmic and biospheric evolution. For the noosphere and noogenics to effectively operate, man must be effectively motivated. Beauty in both its senses, as an esthetic phenomenon of wholeness, radiance and harmony, and as an ethical imperative, achieving the perfection of the good, emanates from the biophilic observations of the naturalist, radiates from the discoveries of physical scientists, and pervades the noosphere with the interest that justifies the immense required labors for its creation and maintenance. Value is a function of interest, and nothing interests man for longer periods or for intenser moments than beauty.

Cosmosphere. The progress of man lies in those

developments, both active and contemplative, which assist life in realizing its destiny in the "cosmosphere", or realm of universal history. The ability to notice our noticing, pay attention to our attention until real choice emerges; then to decide, and to commit the resources to carry out that decision, comprise the requirements to enter the world of history, to become a "vector among the vectors". Part of man's potential is to serve as steward to the biosphere here on Earth, and to assist its spread and evolution through space, perhaps there meeting other biospheres also engaged in the transformation of matter, grand rendezvous which at this moment can only provoke our imaginations with astonishing intimations.

Biosphere II: An Evolutionary Step from Biosphere I

As we have seen, Biosphere I, or its successors limited to planet Earth, has a future at best equal to the life of the Sun in its present form in spite of the biosphere's magnificent adaptability, evolutionary capacity, and geologic power to alter the equilibrosphere of air, water, and rock. Biosphere I must disappear sooner or later unless it can participate in sending forth offspring biospheres to populate other regions of the cosmos.

The creation of new biospheres, energetically and informationally-open, with controlled material access and egress, from the genetic riches of Biosphere I present the opening chapters of a dramatic evolutionary leap for Earthlife. The potential benefits from such creations will accrue both to the remaining career of Earth's biosphere and to life's expansion and evolution throughout space-time-energy.

For Biosphere I, Biospheres II and successors will open the possibility for comparing systems operations; the same cycling of elements and atmospheric regulation will occur but on a scale and time-frequency which can more closely be monitored and/or intentionally modified. For the first time, Biosphere I will have another biosphere to "dialogue with". In a sense, because of the accelerating and uncontrolled drive of man's technical development, Biosphere I has been the subject of large-scale, although always partial, human experiments with no "control" biosphere by which to assess the results.Biospheres II, III, IV, ... n (see chapter 3) will be of inestimable value in understanding the operation of our present biosphere.

Complex, stable, wonder-evoking and evolving biospheres will be a necessity if life is to inhabit other parts of space on a permanent basis. Man's initial space outings, transcendent in possibility, face the danger of becoming only rousing adventures to be recalled before the hearth of historic memory. "Ah, those were great days ... " If life is to spread elsewhere in space, and evolve itself and its surroundings though the potentialities inherent in its creation of free energy, this expansion can only be in the form of biospheric systems, which can be thought of as the primary units, the "space seeds".

Many observers see our present moment as poised on the precipice between evolutionary transformation and annihilation -- at least for the human species. Much of our technology has this two-edged aspect: the rockets that can launch the nuclear arsenal are similar to those which propelled our first voyages to space. The human success story as an animal species likewise threatens that we will fall victim to our uncontrolled population expansion and collapse as swiftly as we increased. Fred Hoyle makes the salient observation that with human population totals climbing from 550 million in 1650, 1.3 billion in 1850, 3.6 billion in 1970, and topping 5 billion in 1986, we are not heading for an unprecedented disaster -- we are already living in the midst of it.[6] This pattern is common in the world of life -- it is the "pressure of life" which drives evolution. And, Lynn Margulis notes, "Populations are beyond good and evil. They grow in response to the availability of space, food, and water. When too numerous, organisms either perish or transcend themselves."[7]

Biosphere II Project: Goals. The major motivation behind creating Biosphere II and developing the capacity to create other micro-scale viable biospheric systems is to assist the biosphere to evolve off planet Earth into potential life regions of our solar system, and eventually throughout the galaxy and cosmos. Other motivations reinforce this overriding objective:

1. Experimental biospheres of a scale in which a human group can directly participate in all the vectors of life will rapidly open up the understanding of Biosphere I which contains extraordinary directions, cycles, and spirals of time, space, mass, energy, diversity and change.

Margulis and Dorion Sagan note: "Only with a full scientific exploration of Gaian control mechanisms can we expect to implement self-supporting living habitats in space. If we are ever to design closed ecosystems that replenish their own vital supplies, we must study the natural technology of the Earth." [8] The creation of Biosphere II and its successors will provide unprecedented research laboratories for the science of biospherics.

2. Biospheres are required for the space program to continue even if only so far as the Moon and Mars, and undoubtedly for the long-term feasibility of permanently-manned Earth-orbit space stations.

3. To produce new understandings for artists, philosophers, scientists, explorers and managers to raise their disciplines to new levels of integrated complexity that can help create new cultures to transform Homo sapien sapiens into a creative collaborator with biospheres, rather than leaving him a parasite weakening his host.

4. To create "living art forms" appropriate to the Time-Space Age which celebrate the epic of evolution with heroes of a new kind. Heroes who are champions of life.

Project Design. The Biosphere II Project consists in the engineering, biological stocking, sealing-off, and operating a basically materially-closed, energetically and informationally open, free energy accumulating life system modelled on the essential elements of Biosphere I.

The most critical design criterion in Biosphere II is creating a harmonious synergy of biosphere and technosphere (see figure 1); utilizing state-of-the-art engineering to assist and back-up the operation of the biota of the system in striving to maintain optimal conditions for the biosphere. Architecture and engineering will complement the biomes of Biosphere II: jungle, desert, savannah, ocean, salt marsh, intensive agriculture and human habitat in striving for maximal diversity (see figure

Biosphere II System Schematic

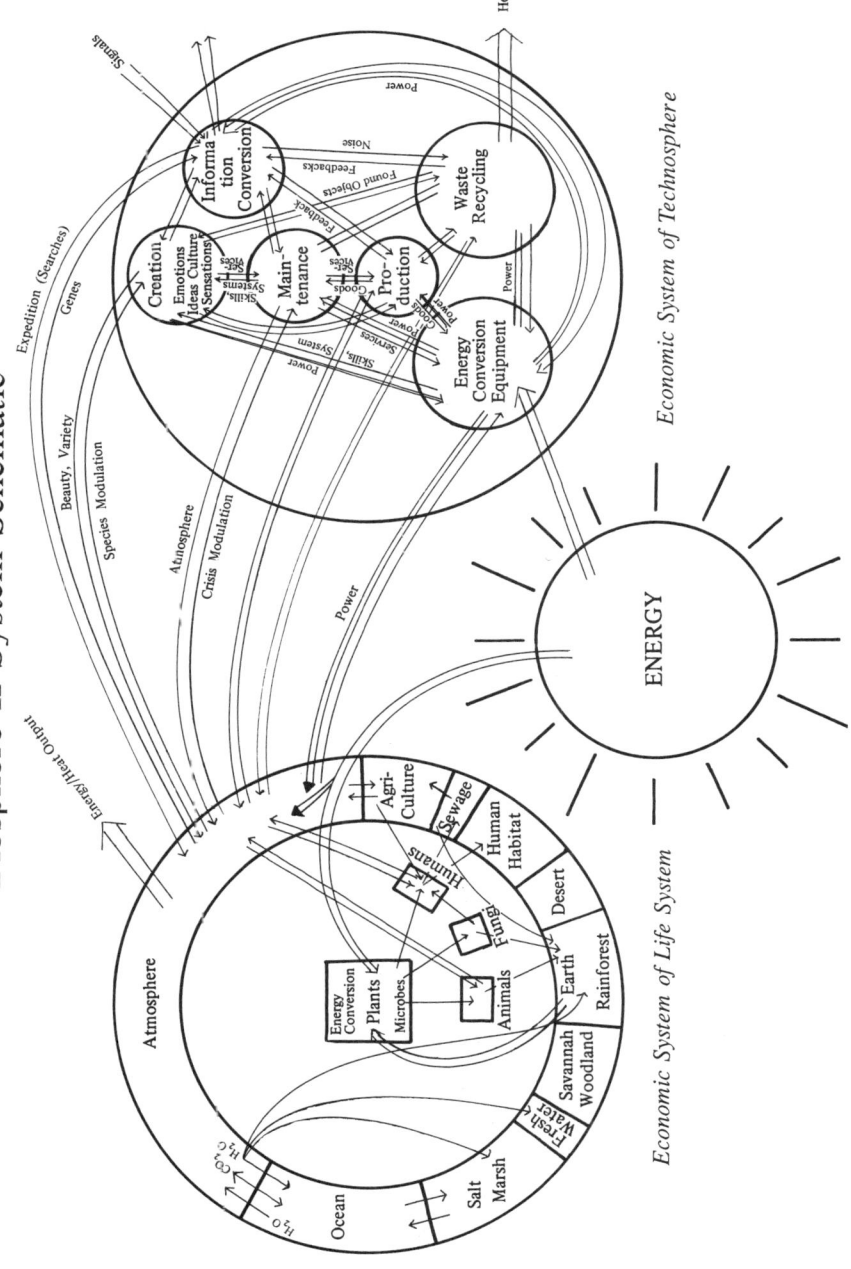

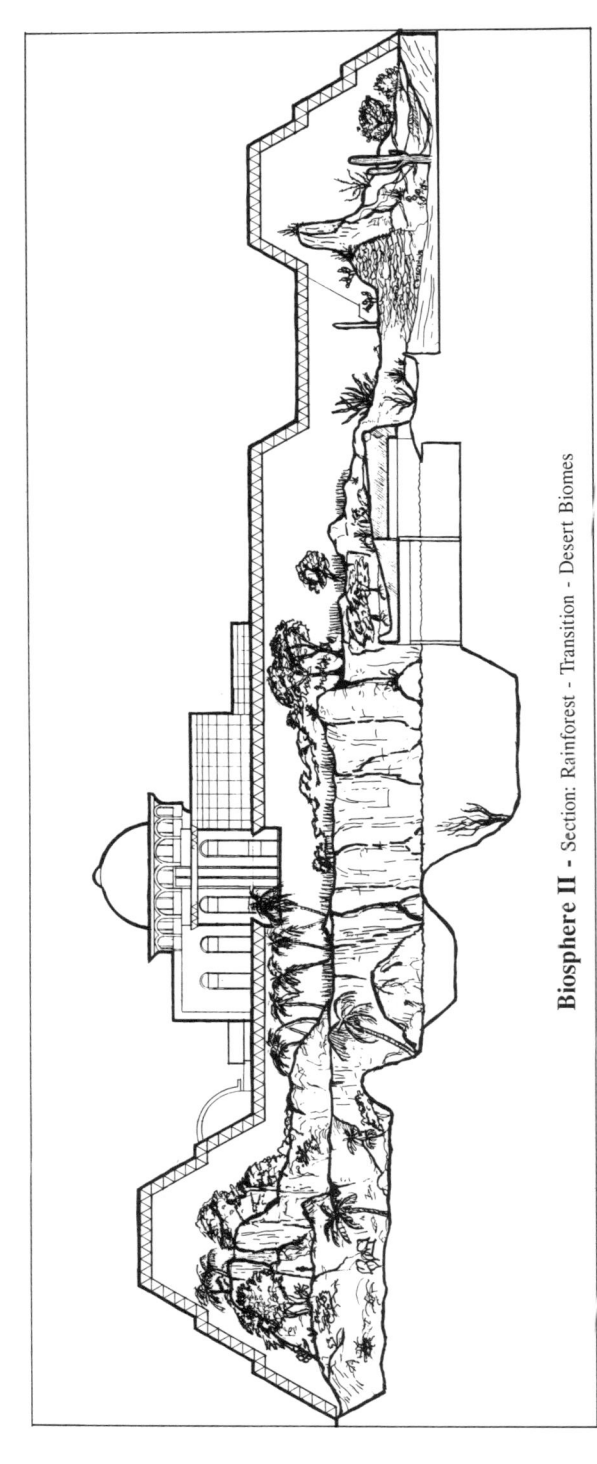

Biosphere II - Section: Rainforest - Transition - Desert Biomes

2), a key safety factor in system stability. Most critically, in Biosphere I both humanity and its industrial works are exponentially increasing while Biosphere II is designed for a maximum human population of 10, a working population of 8, and has its industrial/laboratory areas restricted to definite volume. Expansion is designed to occur modularly, not by risking "'overstocking the petri dish'" with the resulting crash.

Like Biosphere I, Biosphere II will contain rocks, ocean, and air although it will be sealed from below and is so located that volcanic intrusions, earthquake transformations, and metamorphizing pressures from great depths of rock will be absent. However, this will also be true on the Moon and in microgravity and most likely on Mars.

An additional major difference will be that the ocean and landmasses will not be in a 70:30 surface proportion, but rather a 15:85 proportion. However, the ocean in Biosphere II will operate at least at ten times the average productivity of Biosphere I's oceans.

Perhaps the greatest difference will be that the proportion of living biomass to total carbon dioxide in Biosphere II will be on the order of 6,000 times greater than in Biosphere I. This will lead to much more rapid carbon dioxide cycling by the system -- from a period of about eight years in Biosphere I to half a day in Biosphere II. Much evidence has been provided by Clair Folsome's one liter microbial ecosphere experiments at the University of Hawaii that closed systems remain viable due to the adaptive powers of microbial organisms. Oxygen levels tend to be somewhat higher than those in Biosphere I, but stable new rates of atmospheric/water/nutrient cycling are soon achieved. In a certain sense, Biosphere I with its enormous scale "lazes along". The oldest of Folsome's ecospheres has been operating since 1967. None have "died", that is, failed to support dynamic microbial populations.

Biosphere I works in a total approximate equilibrospheric volume of 40 billion cubic miles and Biosphere II in one of 5,000,000 cubic feet. This is a ratio of $1.2 \times 10^{15}:1$. Biosphere II will cover approximately 2 1/2 acres of surface and Biosphere I about 200,000,000 square miles.

A major difference that will occur in both Biosphere I and Biosphere II the moment Biosphere II begins its life-history will be the beginning of objective information about biospheric metabolisms. That is, a comparison of data will immediately commence and underlying principles begin to be inferred from certain phenomena which, superficially different, will reveal at least adumbrations if not patterns of general biospheric laws. All of our information up to now has been flawed by being from only one database, that is, subjective, liable to contain permanent hidden assumptions.

Governing the conception of the Space Biospheres Ventures project are the following six spheres (or pervasives): the cosmosphere, the historic imperative of going into space; the teleosphere, beauty's esthetic of form and beauty's ethic of happiness (the good); noosphere, the necessity of intelligence, micro-incisive and macro-comprehensive to be capable of accomplishing this goal;the technosphere, the capacity to deal with a range of cosmic environments that life without advanced technics could neither travel to nor survive in if it arrived there; the biosphere, the producer of energy available to do work, that gives the surplus value that underlies all the higher possibilities; and the equilibrosphere, the particular thermodynamic system of gas, liquid and solid out of which, when radiant energy impinges sufficiently upon it, life can create the feedback conditions necessary for its growth, maintenance and evolution.

These six basic spheres or pervasives must, then, interpenetrate to make a biosphere capable of moving off the planet Earth.

Project Timetable. Already in operation at the project site, Sunspace Ranch, near Oracle, Arizona are a 17,000 square foot complex housing a prototype recycling intensive agriculture/aquaculture greenhouse facility; a biological tissue culture, laboratory building; solar water heating collectors; and computer/monitoring office. A 12,500 cubic foot test module, its glass atmospherically sealed, has been completed, which is connected to a "lung" of about 5,000 cubic foot capacity to allow for atmospheric pressure expansion/contraction of the structure. This will be used for testing of materials and physiological data, and can later be used for experiments in closed ecological life-support systems or other space station scenarios.

Design and engineering of Biosphere II are scheduled for basic completion the end of 1986, including exact site location and soil testing. Construction, outfitting and final engineering will be done in 1987 and the first half of 1988, and operational experience will be gained through a partial closure stage from mid-1988 through the first half of 1989. August 15, 1989 is the target date for full closure, and Biosphere II will then go on-stream.

Biosphere II is designed to last for a hundred year period

without major repairs, barring major accidents. The knowledge gained and gainable from a biosphere will, of course, increase greatly in value with time since biospheres are adaptive and evolutionary in nature. The first group of eight "biospherians", who are now in training, will operate and live in Biosphere II for a period of two years. An airlock of space suits analogous to what would be used on Mars will provide the way for crew changes to be made.

SBV Tissue Culture Laboratory conducts research and production for the propagation of some of the plants destined for Biosphere II, as well as for present use in the Greenhouse. The lab also conducts research and production activities in the areas of ornamental and agricultural plants for commercial applications, endangered plants, and plants for biological pest control.

BIOSPHERE II

Production Complex: Intensive Agriculture Greenhouse, AquaCulture Bay, architectural drafting studios, and associated support buildings.

Biosphere II Test Module.

REFERENCES

[1] Chaisson, Eric, *Cosmic Dawn*, Berkley Books, New York, 1984, p. 4.
[2] Gribbin, John, *The Death of the Sun*, Delta Books, New York, 1980, p. 62.
[3] Jastrow, Robert, *Red Giants and White Dwarfs*, Harper and Row, New York, 1967, pp. 40-41.
[4] Sagan, Carl, *Cosmos*, p. 234.
[5] Mumford, Lewis, *Technics and Civilization*, Harcourt, Brace, and World, New York, 1934, pp. 4-11.
[6] Hoyle, Fred, *Ten Faces of the Universe*, W.H. Freeman, San Francisco, 1977, pp. 181-195.
[7] Margulis, Sagan, *Microcosmos*, p. 243.
[8] Ibid., p. 275.

3

On the Classification of Biospheres

A biosphere may be defined as a stable, complex, adaptive, and evolving life-system, basically closed to other than local material resources, open to and extracting and storing free energy from radiant energy, open to information, capable of large-scale and comparatively rapid cycle transport and rearrangement of atoms and molecules.

To classify biospheres we will use the following parameters:

--- 1. the five kingdoms of life: monera, protoctista, plant, animal and fungi;
--- 2. the four types of cosmic locations: planets, moons, asteroids, and microgravity space;
--- 3. the three stages: non-technical, local-technical, and space-technical;
--- 4. the two levels: non-biosphere-genic and biosphere-genic.

For example: the biosphere occupying and molding the surface of planet Earth today which can be called *Biosphere I* because it is the first that we know and live in, is classified as a five-kingdom, space-technical, non-biosphere-genic planetary biosphere. However, at one time planet Earth was occupied and its surface molded by a one kingdom (monera), non-technical, non-biosphere-genic, planetary biosphere. That biosphere, type 1, was replaced by type 2: a two kingdom (monera and protoctista), non-technical, non-biosphere-genic, planetary biosphere. Probably type 2 was replaced directly by a five kingdom, non-technical, non-biosphere-genic, planetary biosphere, type 3. About forty thousand years ago type 3 went to type 4 when Earth-bound technics were introduced on a large scale by Homo sapiens sapiens. Less than 30 years ago, space-technics replaced Earth-bound (local) technics, type 6. Type 5 was bypassed on this planet because type 4 did not produce any new biosphere before it produced space-technics. Within three years a type 7 biosphere will arrive: five kingdoms, space-technical, biosphere-genic, planetary biosphere when *Biosphere II* (a type 11 biosphere) emerges on the slopes of the Santa Catalina mountains in Arizona. At the same moment, *Biosphere I* (type 6) will have disappeared and *Biosphere III* (type 7) will have come into being.

The proposed classification looks as follows in its formally complete presentation and will serve both as guide to subject matter and as program for the science of biospherics.

This classification of 49 biospheric types (Table 1) gives a language to deal with all possible biospheres -- presently existing biospheres, past biospheres on planet Earth or elsewhere, biospheres yet to be encountered, and biospheres waiting to be made by biospheres of types 5, 7, 12, 14, 19, 21, 26, 28, 33, 35, 40, 42, 47, or 49.

ON THE CLASSIFICATION OF BIOSPHERES

Types	Kingdoms of Life	Technical Status	Biosphere-Making Capacity	Site
1	1	0	0	PLANET
2	2	0	0	PLANET
3	5	0	0	PLANET
4	5	LT*	0	PLANET
5	5	LT	+	PLANET
6	5	ST*	0	PLANET
7	5	ST	+	PLANET
8	1	0	0	SURPLANET*
9	2	0	0	SURPLANET
10	5	0	0	SURPLANET
11	5	LT	0	SURPLANET
12	5	LT	+	SURPLANET
13	5	ST	0	SURPLANET
14	5	ST	+	SURPLANET
15	1	0	0	MOON
16	2	0	0	MOON
17	5	0	0	MOON
18	5	LT	0	MOON
19	5	LT	+	MOON
20	5	ST	0	MOON
21	5	ST	+	MOON
22	1	0	0	SURMOON
23	2	0	0	SURMOON
24	5	0	0	SURMOON
25	5	LT	0	SURMOON
26	5	LT	+	SURMOON
27	5	ST	0	SURMOON
28	5	ST	+	SURMOON
29	1	0	0	ASTEROID
30	2	0	0	ASTEROID
31	5	0	0	ASTEROID
32	5	LT	0	ASTEROID
33	5	LT	+	ASTEROID
34	5	ST	0	ASTEROID
35	5	ST	+	ASTEROID
36	1	0	0	SURASTEROID
37	2	0	0	SURASTEROID
38	5	0	0	SURASTEROID
39	5	LT	0	SURASTEROID
40	5	LT	+	SURASTEROID
41	5	ST	0	SURASTEROID
42	5	ST	+	SURASTEROID
43	1	0	0	MICROGRAVITY SPACE
44	2	0	0	MICROGRAVITY SPACE
45	5	0	0	MICROGRAVITY SPACE
46	5	LT	0	MICROGRAVITY SPACE
47	5	LT	+	MICROGRAVITY SPACE
48	5	ST	0	MICROGRAVITY SPACE
49	5	ST	+	MICROGRAVITY SPACE

LT = Local Technology
ST = Space Technology
Sur = "On a part of the "

Table One

The classification gives also a means of ranking the biospheres on various scales. For example, it is clear that biospheres of types 7, 14, 21, 28, 35, 42, and 49 are the most complex, and must represent the maximum existential freedom available to life in the cosmos since from their scientific powers each of the other 48 biospheric types could be made by the proper decisions as necessary or as desirable.

On the other hand, it is clear that biospheres of type 1, 8, 15, 22, 29, 36, and 43 are the simplest, being subject to the maximum existential determination by the natural forces of mutation, selection, and material transfer and transformation. However, we see that biosphere type 1 on planet Earth developed more or less unconsciously to a biosphere of type 6, at which point biospheric scientists, managers and engineers can consciously and intentionally move to type 7 and almost certainly forces within type 7 could make members of the next 42 types.

Available for immediate field studies to develop content and depth in the classification system are the historic remains of types 1, 2, 3, and 4 on planet Earth, the present type 6, *Biosphere I,* and the coming to birth of *Biospheres II* and *III,* types 11 and 7. This classification system possess the great power of showing us exactly where to look for data, how to compare, and enables deciding upon a long term program of practical action.

On the Classification of Biospheres, Part II

If the basic notion of the biospheric types is accepted, then we can divide biospheres into two kingdoms, the gravitational (where its material support, corresponding to geosphere in the case of the terrestrial biosphere of the moment, determines its gravity) and non-gravitational or micro-gravity.

So we obtain:

Biospheres

KINGDOMS	Gravitational Biospheres	Micro-gravitational Biospheres
FAMILIES	6	1
TYPES	42	7
KIND	A given individual of a type* and its direct descendants who retain the property of reproducing the original**	Same
VARIETY	A kind modification whose gene banks remain capable of full interchange with those of its parent kinds.	Same

---

* For example, several kinds of individuals of biosphere type

11 would potentially be: tropic, temperate, arctic, oceanic, etc.
** The kinds, of course, can have no descendants except in the cases of types, 5, 7, 12, 14, 19, 21, 26, 28, 33, 35, 40, 42, 47, and 49, because only these can replicate themselves or a new combination of themselves. Therefore all other biosphere kinds will be unique but not changelessly unique because of the possibilities of mutational change to another kind within their type. And of course the biosphere-genic type kinds can have as descendants any of the other type kinds, for example type 7 could "birth" 48 other type-kinds.

4

Mars Settlement

John F. Kennedy immortalized himself and the American space program of the 1960's when he directed, "By the end of this decade we will land a man on the moon and return him safely to Earth." In the intervening time man's capability for space exploration has increased tremendously -- Russian cosmonauts have spent up to eight months living in micro-gravity; spaceships have successfully docked in Earth orbit; robot probes have landed on the surfaces of Mars and Venus; fly-by missions have investigated Saturn, Jupiter and Uranus; satellites have been retrieved and repaired in space by astronauts and remote-control lifting arms; construction techniques have been tested and materials processing has been done in orbit; a re-usable space shuttle is basically engineered; several space probes were able to make precise rendezvous with Halley's Comet; a permanently-manned space station in Earth orbit is nearing accomplishment by the Russians, and is scheduled for the mid-1990's by NASA; a spaceship from planet Earth has left our solar system and is literally headed for the stars.

But clearly there is something wrong. No second president, and all presidents are ambitious, has realized that he would make his place, the American place, and the human place in history forever by saying, "We will make a settlement on Mars as the first step in extending Earthlife among the stars".

We are living in the greatest age of exploration mankind has known -- planets and moons in our solar system are no longer merely astronomical objects but geological ones. Moon rocks and dust are investigated in laboratories around the world. Parts of Mars and the Moon are about as well mapped topographically as remote regions of this planet.

In the face of this factual basis which is, proportionately to the task, far greater than that which Columbus possessed, why the present lack of overwhelming support for the space program? Is it because its direction is unclear, because unformulated, so that space accomplishments appear to be a random assortment of feats and technical wizardry? When will the public leader of vision arise or be pushed up?

Many in the space community surmise that the settlement of Mars has been the through-line objective in the Soviet space program which has enabled them to move so firmly ahead. Their man of vision was Tsiolkovsky.

Man has now the capacity to go along with the biosphere to space, live there, prosper there, and feed back to planet Earth knowledge whose value both to action and contemplation can scarcely be imagined. In America certainly all that is lacking is the political will to say, "We are going to Mars". America possesses the abilities, the know-how, the wealth, the frontier tradition, and the generosity to welcome the future.

Why Mars First? The future of man and biospheres will expand in time to areas throughout our Solar system in micro-gravity orbit, and on the surfaces of planets, moons, and asteroids just as in Biosphere I life seeks ever to expand and fill new econiches. The special importance of establishing biospheres and permanent settlements on Mars is that it will mark an expansion of life to an area sufficiently distanced from the "mother planet", and so rich in territory, resources and opportunity, that a second center of permanent significance for life in our solar system will be established. This certainly would not be the case for settlement in Earth orbit, or even on the Moon, where everything will doubtlessly relate back to Earth-based economics, politics and even culture.

The union of our present astronautics with what we will know about biospherics from Biosphere II means that we will be ready by 1992 to plan, outfit, and set in motion an expedition that can succeed in settlement. What we know already of Mars beckons us to come.

Mars Resources. Mars is the only other planet we know of in our solar system whose surface geology was significantly modified by the action of running water. Whether this means Mars had an early atmosphere similar to Earth's is not known; but even now it has sufficient resources of frozen water -- recent estimates have suggested that if melted, subsoil water deposits could form an "ocean" one thousand feet deep over its surface.

Far from being completely alien, photographs of wind and water-eroded landscapes on Mars cannot be distinguished from photographs of core desert regions on Earth. Its surface is not that different, though a portion of the planet is marked by impact craters dating back to the early solar system.

With an equatorial diameter of 4,170 miles (about one-half that of Earth), Mars has a surface area one-fourth that of Earth, but about the same land area as the total of our continents. Its average density is some four grams/square centimeter, about 40 percent lower than Earth. Martian surface gravity is about 38 percent that of the terran, but its atmosphere, principally carbon dioxide, is less than one percent of the pressure of Earth's. Reasons for this anomaly are not completely clear.

The gravity of Mars offers significant advantages for Earth-evolved animal and plant life which have certain problems with the micro-gravity of space. At what level of hypogravity the lack of root orientation experienced by plants, and the loss of vertebrate bone calcium ceases to be dysfunctional is unknown, but gravity reduced to the Martian level may well have enhanced physiological effects on plant and animal physiology -- less energy will be needed to fight the gravity making everything on Earth's surface fall at 32 feet per second per second. Moving out from the Martian gravity well for further space expeditions will be, of course, far easier (11,214 miles per hour escape velocity compared to Earth's 25,000 miles per hour).

The Martian day-length of 24 hours 37 minutes probably means circadian Earth rhythms will suit life on Mars well. The Martian year, however, is equivalent to 687 Earth days.

Mars has a planetary rotation axis tilted very nearly the same as Earth's, producing similar seasonal changes. Mars has polar caps covered with ice, mainly frozen carbon dioxide, which can extend to latitudes of 60 degrees in the

North and 45 degrees in the South. Because of Mars' elliptical orbit its southern hemisphere has shorter but hotter summers than the northern, so that its carbon dioxide icecap virtually disappears there, sublimating directly into the gaseous state. It is thought that water constitutes one portion of the polar ice caps which never melts. Billions of tons of water are contained in the polar caps.

Average temperature on Mars is -23 degrees C (-10 degrees F). Summer daytime temperatures in the equatorial region of Mars can reach 45 degrees F. The scant atmospheric mass of Mars generates much larger diurnal and equator/pole variations in temperature than the Earth, which helps generate its winds and duststorms. However, the Martian storms do not completely occlude sunlight and serve a useful function by raising atmospheric temperatures as much as 25 degrees C in a single day. They can continue for months. The Viking weather monitoring revealed a welcome surprise: the Martian duststorms are not violent near the surface, at five feet they are 6 to 22 feet/second or 5 to 16 miles/hour.[1]

Little is definitely known about Martian ore-bodies, though its planetary similarity in size and position to Earth suggests there will be significant findings despite the apparent lack of tectonic plate movements. Mars possesses "crustal swells, large rifts, and certain types of volcanoes often linked with mineral deposits on Earth ... interplanetary analogs strongly suggest the possibility of economically significant Martian mineral caches ... in particular, mineral-rich Africa seems to share many volcanic and tectonic characteristics with portions of Mars and may be suggestive of the potential mineral wealth of Mars".[2]

Over twenty elements have already been identified in the Martian atmosphere and surface. Oxygen is present as a trace gas in the atmosphere and is quite abundant in the iron oxide compounds which color the "Red Planet". Nitrogen is present at about 2.7 percent of the atmosphere. Resources are available for making rocket propellant and energy fuels from Martian materials -- some of the combinations being considered are hydrogen peroxide, methane/oxygen, carbon monoxide/oxygen.

Terraforming/Biosphering. There has been renewed interest recently in the possibilities for "terraforming" Mars -- altering its atmospheric density by the introduction of greenhouse gases and so making surface temperatures more hospitable. This would also liberate quantities of frozen water, from Martian regolith and icecap, and reduce the severity of duststorms.

James Lovelock and Michael Allaby outline in their *Greening of Mars* a delightful scenario in which world disarmament produces a disposal problem for our military rockets, which loaded with industrial pollutant chlorofluorocarbons, can be dispatched to Mars -- along with an assorted CARE package of eager microbes. From what we know of Martian conditions, it would not take a great shift in atmosphere/climate to enable our ever-resourceful microbial forebears to get a foot-hold. Carl Sagan has suggested alterations of the planet's albedo by establishment of black bacteria on the polar caps, or even the use of carbon black.

These and other terraforming ideas, if verified to be feasible, could complement well the practical approach of placing and replicating enclosed biospheres on the surface of bases while Martian resources and history are investigated, as places to live and prosper, and as biological laboratories to study and accelerate the evolution of life to adapt to Martian conditions. Simultaneously, of course, efforts could proceed in modifying planetary conditions themselves, which even if not able to proceed to a terraformed condition, would still mean a more tolerant equilibrosphere to venture forth in from the biosphere settlements.

With time, the two approaches might converge and we would meet other intelligent life in the universe -- the evolved/adapted Martian progeny of Biosphere I. Earthlife will have another planetlife to "talk to" and "think with"; we will no longer be alone whether on Earth or Mars.

Where to Go. It was perhaps historically unfortunate that the Viking landings of the 1970's were made in such a visually uninspiring part of the Red Planet due to fears about safe touch-down -- much as if an extraterrestrial's first view of Earth was a stretch of the flat Nevada desert rather than the Grand Canyon area. Had they landed in the Coprates Canyon (Valles Marineris) we would have looked at an ancient tributary grand canyon system which dwarfs our own -- it stretches over 2,500 miles, is 120 miles wide and two miles deep. Another breath-taking location would be near Olympus Mons -- the largest known mountain in the solar system -- a volcano over 400 miles across and 15 miles high. That makes it almost three times the height of Mount Everest and about the size of Colorado!

Another candidate region for the first Martian biospheres is the Hellas Basin, a 1,000 mile diameter basin, in the southern highlands. Probably covered with volcanic basaltic lava several hundred million years after basin formation,

Hellas has a comparatively regular surface topography. Impact and volcanic craters constitute some 40 percent of the equatorial region and dominate the southern hemisphere of Mars -- some 50 thousand craters exceeding one kilometer in diameter have been identified. One attraction is that a low-lying and therefore heavier atmospheric basin such as Hellas has conditions able to support water in its liquid phase.

Biospheres in the Mars Settlement. The following is a practical scenario for the operation of a permanent Martian settlement:

Biosphere II will provide both the model and data for its improvement for the fundamental module that will allow the successful building and operation of the Mars settlement.

Four radiation-proof biospheres, each designed for full biological and life/esthetic support for six to ten people will be arranged in four directions around a central technical center with a reserve "ocean" to moderate the environment and provide an extra life-support system (see illustrations 1 and 2). This reserve ocean and its pleasant shores will not, however, be designed as a full biosphere because this "commons area" will be open for full exchanges with material from Mars and the Earth's technosphere, and this part of the settlement will be the center for the community's technosphere and plaza for public meetings. The technical center will allow the biological, transport, mining, and operations groups their work areas, as well as providing a Hospitality Center for visitors from Earth, moon, microgravity, or other parts of Mars.

The overall volume of Mars Settlement Stage I is estimated at 40,000,000 cubic feet for the four interconnected biospheres. All of the biospheres would be

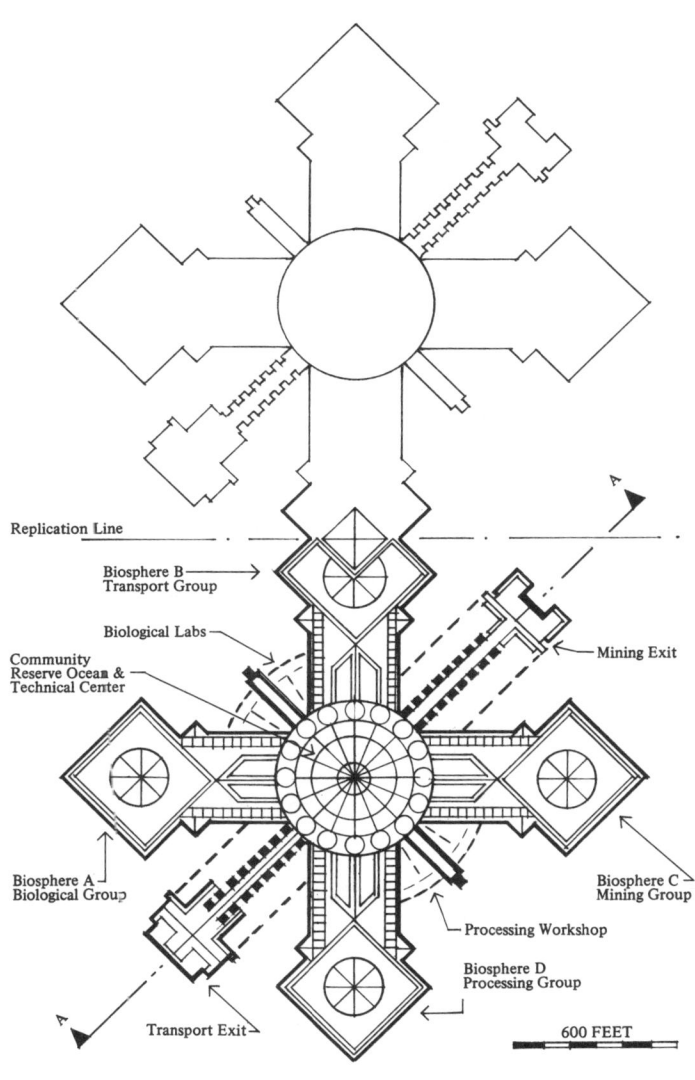

Plan

Mars Science Station and Hospitality Base

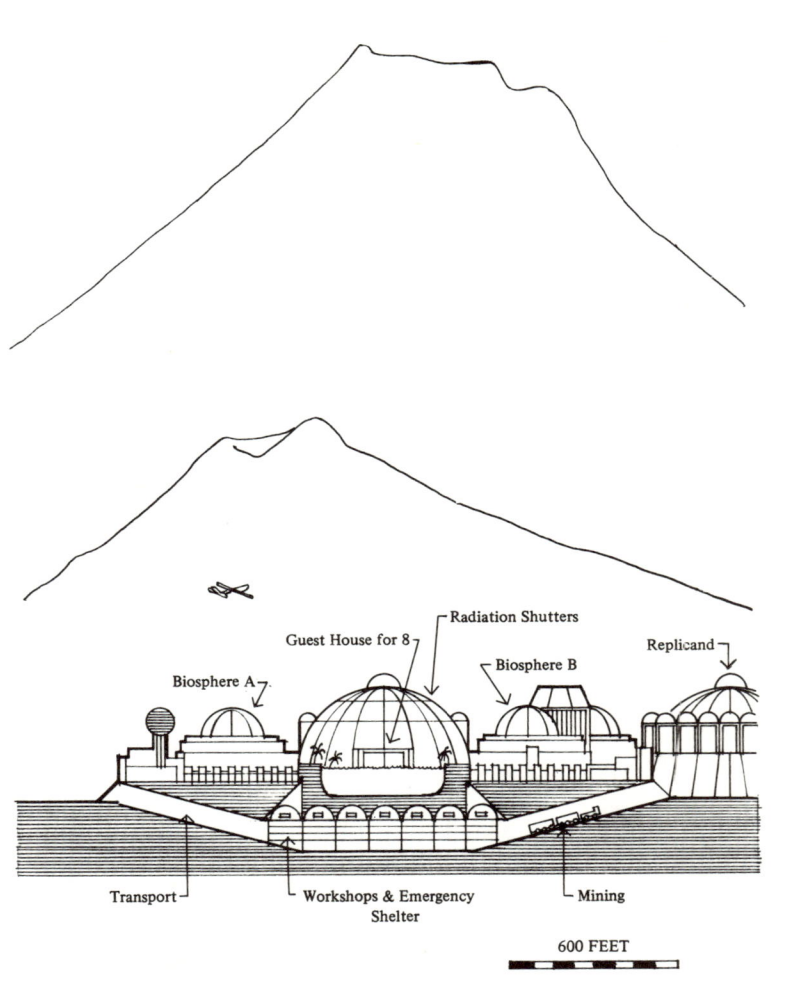

Section A - A
Mars Science Station and Hospitality Base

stocked with carbon dioxide, water and oxygen derived mechanically from the nearby area, but once operation of the biosphere commences further cycling and increase of these elements will be accomplished by microbe/fungal/plant/animal metabolism just as on Earth. Mars Settlement I will not be sealed below from the Martian crust and the action of the biospheres will be to incorporate new material into the biotic cycle, and to create free energy -- the transformation of the surface of Mars by life will have commenced.

Biospheres A, B, C, and D will be built and not "one big biosphere" for safety in case of plague, impact breakages, etc. The "biologist's group" will concentrate in the areas of ecology, medicine, and agriculture. The transport team will especially focus on astronautics, machine repair and construction, and in protective coverings. The mining group will target geology, extractive, concentrating and refining operations of solids, liquids, and gases, and in radioactivity. The processing task group will handle chemical engineering, electronics, and manufacturing.

Each group will assist the others in the accomplishment of their individual tasks and in the overall objective: to create and master the art of living on Mars.

Each individual member regardless of group shall have access to the electronic communication system (and be entitled to free use of this system.)

A weekly half-day reserve ocean rendezvous will be held for face-to-face contact of the entire community. Contact will be informal.

The Strategic Command will consist of one member from each group. This group will serve for four years and elect its chairman annually. At the end of four years each group will again choose its member for the next four year term.

A member of the sex opposite to the chairman will be chosen by each group to serve as secretary to its Strategic Command member. The secretary's duty will be to report to each group what occurred at the meetings. Meetings will operate under *Robert's Rules of Order* protocol. The secretaries are to act only as observers during the meetings. They will be chosen for two terms. A technical standing committee will be chosen, one from each group, that can be called upon for advice by the Strategic Command.

Each group shall prepare a monthly report of its activities, and the Strategic Command shall make a monthly overall assessment of the task. One rendezvous each month will have these reports made to the entire community. Four mixed groups (that is, two from each primary group if there are eight people in each) will discuss the reports in order to broaden and deepen understanding of the reports' content.

The first Mars Base will at first finance itself by contracts for information, artistic works, development of processes for living on Mars to be given out to any UN, USA, USSR or other space nation approved mission upon payment of development fees at an agreed rate, and for use of the Hospitality Center and purchase of foods.

Mars Settlement I is designed to expand into Mars Settlement II which will be basically a duplicate. The two communities will meet at alternate reserve oceans on alternate months where the two Strategic Commands will form an Objectives/Policies Board to evaluate risks and opportunities which emerge and monitor progress towards agreed goals.

There will be a semi-annual day-long celebration to be held at alternate reserve oceans. The population can range from 64 to 80 people. If more population arrives they will have to begin their own communities, which will be increasingly be made of indigenous Martian materials as the technical infrastructure develops.

The Mars Settlement will commit itself to remain non-military, and to base its rules upon the findings of, and in the spirit of, the use of the scientific method. No one will be accepted without a thorough grounding in science and people born into the community will be taught how to become citizens of the world of science as their essential birthright. A critical intellect, taught how to observe, and act as needed will be a survival necessity -- dogma or ideology will only hamper the reality-thinking required of space pioneers.

The Mars settlement will seek its security, growth, and evolution in the service of providing truths about the nature of the universe in its region to all interested parties, and to becoming a center for an invaluable history of the planet Mars and at least part of the cutting edge of history of Homo sapiens sapiens.

REFERENCES

[1] Oberg, James E., *New Earths*, New American Library, New York, 1983, pp. 178-179.

[2] Cordell, Bruce, "A Preliminary Assessment of Martian Natural Resource Potential", *A Case For Mars II*, McKay, Christopher, ed., American Astronomical Society, San Diego, 1985, pp. 633-4.

5

Challenges and Opportunities

The fate of man, which we daily endure, is to be lost in time, separated in space, entropic in energy and information, limited in matter, deficient in data, uncertain in perception, and certain only of continuing folly.

Man's destiny, however, is the conquest of time, the liberation of space, redundancy of energy and information, asymptotic to unlimited in matter, asymptotic to objective universal data in the human, biospheric, and cosmic realms, trained perception conscious of its limits, and the attaining of wisdom.

Time. Man could be lost in a few million year long segment of time in a universe of tens of billions of years of time, but man could co-exist with the universe with the proper spread of biospheres and thus achieve cosmic immortality.

Space. Man could be separated in space to the surface of one small peripheral planet and its neighboring moon; or man could move freely (within the limitations of mastery of physical laws) throughout the billions of light-years of space.

Energy and Information. Man can run out of useful energy and information due to the limitations of the Sun, or by the loss of freedom and intelligence in human culture; however, by gaining myriad access to stellar radiation and to those realms of thoughtful experiments called art, science, management, adventure, philosophy, and the transcendent, man could gain such redundancy as to enable the overcoming of entropy (the disorganization of order and the ability to do work) within the limits of any remaining energy radiating source in the universe from which free energy can be extracted.

Data and Perception. The proper solutions to the above lead to new horizons of perception and new inputs of patterns of data on a truly total systems scale, whereas now man lacks both data and perception about the true nature of humanity. What would humanity be like under the many different gravitational, electromagnetic, social, psychological, technical conditions beyond anything Earth can offer? And with these different systems in networked scientific communication?

Folly and Wisdom. The certainty of folly lies in the complacent self-importance inescapable when man takes relative ignorance to represent knowledge or even truth and refuses to seek arduously, skillfully, and to the ends of the cosmos for knowledge. This commitment to the search would give man the possibility of wisdom. Man should not settle for anything less. Space biospheres will be, with astronautics, one of the two tools rendering such wisdom achievable to those cultures which will accept nothing less than succeed.

Bibliography and Suggested Reading

Billingham, John (ed.), *Life in the Universe*. MIT Press, Cambridge, Mass., 1981.

Boston, Penelope J. (ed.), *The Case for Mars*. American Astronautical Society, San Diego, 1984.

Chaisson, Eric, *Cosmic Dawn*. Berkley Books, N.Y., 1984.

Crick, Francis, *Life Itself*. Simon and Schuster, N.Y., 1981

Darwin, Charles, *The Origin of Species*. 6th Ed. Oxford Univ. Press, London, 1872.

Delwiche, C.C., "The Nitrogen Cycle" in *The Biosphere*. W.H. Freeman, San Francisco, 1970.

Ehrlich, Paul R., *The Population Bomb*. Sierra Club, 1968.

Fuller, R. Buckminster, *Critical Path*. St. Martin's Press, New York, 1981.

Glasstone, Samuel, *The Book of Mars*. NASA SP-179, Washington D.C., 1968.

Gribbin, John, *The Death of the Sun*. Delta Books, N.Y., 1980.

Hoyle, Fred, *Ten Faces of the Universe*. W.H. Freeman, San Francisco, 1977.

Jastrow, Robert, *Red Giants and White Dwarfs*. Harper and Row, N.Y., 1967.

Kamshilov, M.M., *The Evolution of the Biosphere*. Mir Publishers, Moscow, 1976.

Lapo, A.V., *Traces of Bygone Biospheres*. Mir Publishers, Moscow, 1982.

Lovelock, J.E., *Gaia, A New Look at Life on Earth*. Oxford University Press, Oxford, 1979.

Lovelock, J., and M. Allaby, *The Greening of Mars*. St. Martins, N.Y., 1984.

Margulis, Lynn and Karlene U. Schwartz, *The Five Kingdoms of Life*. W.H. Freeman, San Francisco, 1981.

Margulis, Lynn and Dorion Sagan, *Microcosmos*. Summit Books, N.Y., 1986.

McKay, Christopher (ed.), *The Case for Mars II*. American Astronautical Society, San Diego, 1985.

Milne, D., Raup, D., Billingham, J., Niklaus, K. and K. Padian (ed.), *The Evolution of Complex and Higher Organisms*. NASA SP-478, Washington, D.C., 1985.

Bibliography and Suggested Reading

Mitten, M.A. (ed.), *The Cambridge Encyclopedia of Astronomy.* Crown Publishers, N.Y., 1977.

Mumford, Lewis, *Technics and Civilization.* Harcourt, Brace and World, N.Y., 1934.

Myers, Norman (ed.), *Gaia Atlas of Planetary Management.* Pan Books, London, 1985.

Oberg, James E., *New Earths.* New American Library, N.Y., 1983.

Odum, Eugene P., *Ecology.* Second Edition. Holt, Rinehart and Winston, London, 1975.

O'Neill, Gerard K., *The High Frontier.* Doubleday, N.Y., 1982.

Pioneering the Space Frontier. The Report of the National Commission on Space. Bantam, N.Y., 1986.

Raven, P., Evert, R., and Eichhorn, S., *The Biology of Plants.* Worth Publishers, N.Y., 1986.

Sagan, Carl and I.S. Shklovskii, *Intelligent Life in the Universe.* Delta Books, N.Y., 1966.

Sagan, Carl, *Cosmos.* Random House, N.Y., 1984.

Sevastyanov, V.I., Ursul, A.D., and Y.A. Shkolenko, *The Universe and Civilization.* Progress Publishers, Moscow, 1981.

Snyder, T.P. and John Allen (ed.), *The Biosphere Catalogue.* Synergetic Press, London, 1985.

Soviet Geography Today: Aspects of Theory. Progress Publishers, Moscow, 1981.

Soviet Geography Today: Physical Geography. Progress Publishers, Moscow, 1982.

Taylor, Theodore B. and Charles C. Humpstone, *The Restoration of The Earth.* Harper and Row, New York, 1973.

Vernadsky, Vladimir, *The Biosphere.* Synergetic Press, London, 1986.

Weinberg, Steven, *The First Three Minutes.* Bantam, N.Y., 1979.

Wilson, Edward O., *Biophilia.* Harvard University Press, Cambridge, Mass., 1984.

Other biospheric publications from Synergetic Press

The Biosphere
by V.I. Vernadsky

The first English edition of the classic work by Vernadsky originally published in Russian in 1926.

The Biosphere Catalogue
Editor in Chief, T.P. Snyder
Scientific Editor, John Allen

A comprehensive presentation of the biosphere, with contributions from over thirty leading figures in fields ranging from atmosphere, hydrosphere, geosphere, plants and animals to cultures, cities, space biospheres, genetics and travel.

Feng-Shui
by Ernest J. Eitel
with commentary by John Michell

The science of sacred landscape in old China. The first English treatise ever written on the Chinese code of practice used in overall matters of architectural design, city planning and use of the countryside.